William S. Franklin, Robert Baird Williamson

The Elements of alternating Currents

William S. Franklin, Robert Baird Williamson

The Elements of alternating Currents

ISBN/EAN: 9783743343399

Manufactured in Europe, USA, Canada, Australia, Japa

Cover: Foto ©ninafisch / pixelio.de

Manufactured and distributed by brebook publishing software (www.brebook.com)

William S. Franklin, Robert Baird Williamson

The Elements of alternating Currents

THE
ELEMENTS OF ALTERNATING CURRENTS

THE

ELEMENTS

OF

ALTERNATING CURRENTS

BY

W. S. FRANKLIN

AND

R. B. WILLIAMSON

New York
THE MACMILLAN COMPANY
LONDON: MACMILLAN & CO., Ltd.
1899

All rights reserved.

PREFACE.

THIS book represents the experience of seven years' teaching of alternating currents, and almost every chapter has been subjected repeatedly to the test of class-room use. The authors have endeavored to include in the text only those things which contribute to the fundamental understanding of the subject and those things which are of importance in the engineering practice of to-day.

It may be taken for granted that the authors are deeply indebted to Mr. C. P. Steinmetz, whose papers are unique in their close touch with engineering realities.   W. S. F.

SOUTH BETHLEHEM,
    June, 1899.

TABLE OF CONTENTS.

CHAPTER I.
                                                                         PAGE.

Magnetic flux. Induced electromotive force. Inductance. Capacity............. 1

CHAPTER II.
The simple alternator. Alternating e. m. f. and current. The contact maker.... 18

CHAPTER III.
Measurements in alternating currents. Ammeters. Voltmeters. Wattmeters.... 29

CHAPTER IV.
Harmonic electromotive force and current... 39

CHAPTER V.
Problem of the inductive circuit. Problem of the inductive circuit containing a condenser. Electrical Resonance.. 51

CHAPTER VI.
The use of complex quantity.. 62

CHAPTER VII.
The problem of coils in series. The problem of coils in parallel. The problem of the transformer without iron.. 69

CHAPTER VIII.
Polyphase alternators. Polyphase systems.. 80

CHAPTER IX.
The theory of the alternator. Alternator designing.. 95

CHAPTER X.
The theory of the transformer.. 119

CHAPTER XI.
Transformer losses and efficiency. Transformer connections. Transformer designing... 137

CHAPTER XII.
The synchronous motor... 151

CHAPTER XIII.
The rotary converter... 166

CHAPTER XIV.
The induction motor.. 178

CHAPTER XV.
Transmission lines... 198

SYMBOLS.

i instantaneous value of current.
$\boldsymbol{I}$ maximum value of an harmonic alternating current.
I effective value of an alternating current.
e instantaneous value of e. m. f.
$\boldsymbol{E}$ maximum value of an harmonic alternating e. m. f.
E effective value of an alternating e. m. f.
r R resistance (r sometimes used for radius).
L inductance.
J electrostatic capacity.
t time.
T Z turns of wire.
z turns of wire per unit length of a coil.
n speed in revolutions per second.
f frequency in cycles per second.
ω frequency in radians per second.
μ magnetic permeability.
l length.
q sectional area.
N magnetic flux.
B flux density.

THE ELEMENTS OF ALTERNATING CURRENTS.

CHAPTER I.

INDUCTANCE AND CAPACITY.

1. Magnetic flux.—Let a be an area at right angles to the velocity of a moving fluid, and let v be the velocity of the fluid. Then av is the flux of fluid across the area in units volume per second. Similarly the product of the intensity, f, of a magnetic field into an area a at right angles to f is called the magnetic flux across the area. That is

$$N = fa \qquad (1)$$

in which N is the magnetic flux across an area a which is at right angles to a magnetic field of intensity f.

The unit of magnetic flux is the flux across one square centimeter of area at right angles to a magnetic field of unit intensity. This unit flux is called a *line of force** or simply a *line*. For example, the intensity of the magnetic field in the air gap between the pole face of a dynamo and the armature core is, say, 5000 units, and this field is normal to the pole face of which the area is 300 square centimeters, so that 1,500,000 lines of magnetic flux pass from the pole face into the armature core.

The trend of the lines of force near the poles of a magnet is shown in Fig. 1. In Fig. 2 is shown the trend of the lines of force through a coil of wire in which an electric current is flowing.

---

* A line of force is a line drawn in a magnetic field so as to be in the direction of the field at each point. The term *line of force* is used for the unit flux for the following reason: Consider a magnetic field. Imagine a surface drawn across this field. Suppose this surface to be divided into *parts* across each of which there is unit flux. Imagine lines of force drawn in the magnetic field so that one line of force passes through each of the *parts* of our surface. Then the magnetic flux across any area anywhere in the field will be equal to the *number of these lines* which cross the area.

Magnetic flux through a coil.—In the discussion of the inductances of coils, it is customary to speak of the *magnetic flux through a coil* as the product of the number of lines through the

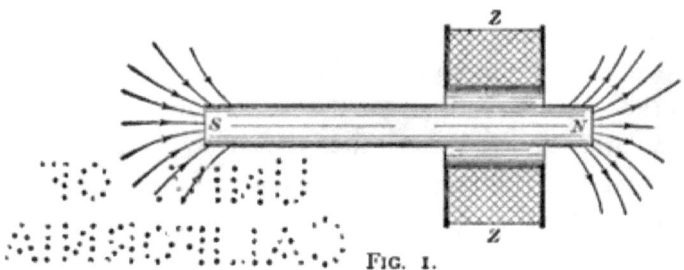

FIG. 1.

opening * of the coil into the number of turns of wire in the coil; that is, the lines are counted as many times as there are turns of wire.

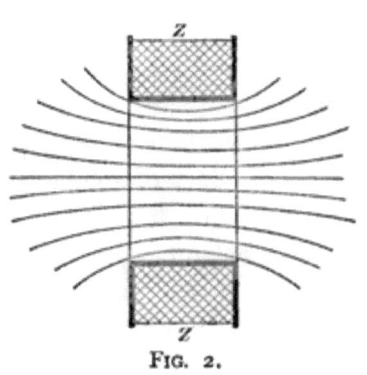

FIG. 2.

2. Induced electromotive force.—When a bundle of Z wires connected in series moves across a magnetic field so as to cut the lines of force, in each wire an e. m. f. is induced which is equal to the rate, $\frac{dN}{dt}$, at which lines of force are cut, and the total e. m. f. induced in the bundle of wires is

$$e = -Z\frac{dN}{dt} \qquad \text{(ii)}$$

Similarly when the magnetic flux through the opening of a coil changes an e. m. f. is induced in the coil, such that

$$e = -Z\frac{dN}{dt} \qquad \text{(ii) bis}$$

in which Z is the number of turns of wire in the coil, $\frac{dN}{dt}$ is the

---

* Strictly, the number of lines through a mean turn multiplied by the number of turns.

rate of change of the flux, and e is the induced e. m. f. The negative sign is chosen for the reason that an increasing positive flux produces a left-handed e. m. f. in the coil.*

Examples.—(*a*) A conductor on a dynamo armature cuts the 1,500,000 lines of force from one pole face in, say, $\frac{1}{50}$ second, that is, at the rate of 75,000,000 lines per second; and this is the e. m. f. (in c. g. s. units) induced in the conductor.

(*b*) A coil Z, Fig. 1, having Z turns of wire surrounds a magnet NS through which there are N lines of flux. The coil is quickly removed from the magnet, reversed, and replaced; the whole operation being accomplished in t seconds. The flux N being reversed with respect to the coil is to be considered as changing $+N$ to $-N$, the total change being therefore $2N$. Dividing this total change of flux by the time t gives $\frac{2N}{t}$, which is the average value of $\frac{dN}{dt}$ during the time t, and the average e. m. f. induced in the coil is $\frac{2N}{t} \cdot Z$. This e. m. f. is expressed in c. g. s. units and is to be divided by 10^8 to reduce it to volts.

3. The magnetic field as a seat of kinetic energy.—The magnetic field is a kind of obscure motion of the all pervading medium, the ether; and this motion represents energy. The amount of energy in a given portion of a magnetic field is proportional to the square of the intensity of the field. This is analogous to the fact that the kinetic energy of a portion of a moving liquid is proportional to the square of the velocity of the liquid.

4. Kinetic energy of the electric current in a coil. *Definition of Inductance.*—The kinetic energy of an electric current is the energy which resides in the magnetic field produced by the current. The kinetic energy is, at each point, proportional to the square of the field intensity, that is, to the square of the current.

---

* This, although an inadequate statement, must suffice; especially inasmuch as the sign in equation (ii) is of no practical importance.

Therefore the total kinetic energy of the field is proportional to the square of the current. That is,

$$W = \tfrac{1}{2} L i^2 \qquad (1)$$

in which W is the total energy of a current i in a given coil, and $(\tfrac{1}{2}L)$ is the proportionality factor. The quantity L is called the *inductance* of the given coil.

Units of Inductance.—When in equation (1) W is expressed in joules and i in amperes, then L is expressed in terms of a unit called the *henry*. When W is expressed in ergs and i in c. g. s. units of current, then L is expressed in c. g. s. units of inductance. The c. g. s. unit of inductance is called the *centimeter*, for the reason that the square of a current must be multiplied by a length to give energy or work; that is, inductance is expressed as a length and the unit of inductance is, of course, the unit of length. The henry is equal to 10^9 centimeters of inductance.

Example: A given coil with a current of 0.8 c. g. s. units produces a magnetic field of which the total energy is 6,400,000 ergs, so that the value of L for this coil is 20,000,000 centimeters. If the current is expressed in amperes and energy in joules then the total energy corresponding to 8 amperes would be 0.64 joules and the value of L would be 0.02 henry.

Non-inductive circuits: A circuit of which the inductance is negligibly small is called a *non-inductive circuit*. Since the inductance of a circuit depends upon the energy of the magnetic

Fig. 3.

field, therefore a non-inductive circuit is one which produces only a weak field, or a field which is confined to a very small region. Thus, the two wires, Fig. 3, constitute a non-inductive circuit, especially if they are near together; for these two wires with opposite currents produce only a very feeble magnetic field in the surrounding region. The wires used in resistance boxes are usually arranged non-inductively. This may be done by doubling the wire back on itself and winding this double wire on a spool. In this case the e. m. f. between adjacent wires may be great and they may have considerable electrostatic capacity. In order to make a non-inductive resistance coil without this defect, the wire may be wound, in one layer, on a thin paper cylinder so as to bring the terminals as far apart as possible. This cylindrical coil is then flattened so as to reduce the region (inside) in which the magnetic field is intense. This gives a non-inductive coil of which the electrostatic capacity is inconsiderable.

5. Moment of inertia, analogue of inductance.—The kinetic energy of a rotating wheel resides in the various moving particles of the wheel. The velocity (linear) of each particle of the wheel is proportional to the speed (angular velocity) of the wheel, and the energy of each particle is proportional to the square of its velocity, that is, to the square of the speed. Therefore, the total kinetic energy of the wheel is proportional to the square of the speed. That is,

$$W = \tfrac{1}{2} K \omega^2 \qquad (2)$$

in which W is the total energy of a wheel rotating at angular velocity ω and $(\tfrac{1}{2}K)$ is the proportionality factor. The quantity K is called the *moment of inertia* of the wheel.

6. Proposition.—*The inductance of a coil wound on a given spool is proportional to the square of the number of turns, Z, of wire.* For example, a given spool wound with No. 16 wire has 500 turns and an inductance of say, 0.025 henry; the same spool wound with No. 28 wire would have about ten times as many turns and its inductance would be about 100 times as great or 2.5 henrys.

Proof: To double the number of turns on a given spool would everywhere double the field intensity for the same current, and therefore the energy of the field would everywhere be quadrupled for a given current so that the inductance would be quadrupled according to equation (1).

7. Proposition.—*The inductance of a coil of given shape is proportional to its linear dimensions, the number of turns of wire being unchanged.* For example, a given coil has an inductance of 0.022 henry, and a coil three times as large in length, diameter, etc., has an inductance of 0.066 henry.

8. Electro-motive force required to make a current in a coil change.—A current once established in a coil of zero resistance would continue to flow without the help of an e. m. f. to maintain it just as a wheel when once started continues to turn, provided there is no resistance to the motion of the wheel. To increase the speed of the wheel a torque must act upon it in the direction of its rotation, and to increase the current in the coil an e. m. f. must act on the coil in the direction of the current.

When an e. m. f. e (over and above the e. m. f. required to overcome the resistance of the coil) acts upon a coil the current is made to increase at a definite *rate*, $\frac{di}{dt}$, such that

$$e = L\frac{di}{dt} \tag{3}a$$

Proof of equation (3): Multiplying both members of this equation by the current i we have $ei = Li\frac{di}{dt}$. Now ei is the rate, $\frac{dW}{dt}$, at which work is done on the coil in addition to the work used to overcome resistance, and this must be equal to the rate at which the kinetic energy of the current in the coil increases. Differentiating equation (1) we have $\frac{dW}{dt} = Li\frac{di}{dt}$. Therefore, equation (3) is proven.

Torque required to made the speed of a wheel increase: When a torque, T (over and above the torque required to overcome the frictional resistance), acts upon a wheel, then the angular velocity, ω, of the wheel is made to increase at a definite *rate* $\frac{d\omega}{dt}$ such that

$$T = K\frac{d\omega}{dt} \tag{4}$$

Proof of equation (4): Multiplying both members of this equation by the angular velocity, ω, of the wheel we have $T\omega = K\omega\frac{d\omega}{dt}$. Now $T\omega$ is the rate $\frac{dW}{dt}$ at which work is done on the wheel and this must be equal to the rate at which the kinetic energy of the wheel increases. Differentiating equation (2) we have $\frac{dW}{dt} = K\omega \frac{d\omega}{dt}$. Therefore equation (4) is proven.

9. Magnetic flux through a coil due to a current in the coil.—In dealing with coils it is usual to speak of the *magnetic flux through the coil* as the product of the flux through the opening of the coil or the flux through a mean turn, multiplied by the number of turns as pointed out in Art. 1. That is

$$N = ZN'$$

in which N' is the flux through the opening of a coil (through a mean turn), Z is the number of turns of wire in the coil and N is what is called the *flux through the coil*.

Proposition.—The flux N through a coil due to a current i in in the coil is

$$N = Li \qquad (5)*$$

in which L is the inductance of the coil. This proposition is proven in the next article.

10. Self-induced e. m. f. *Reaction of a changing current.*—When one pushes on a wheel, causing its speed to increase, the wheel reacts and pushes back against the hand. This reacting torque is equal and opposite to the acting torque $K\frac{d\omega}{dt}$ [equation (4)], which is causing the increase of speed. Thus, when the speed of the wheel is increasing, the reacting torque is in a direction opposite to the speed, and, when the speed is decreasing, the reacting torque is in the same direction as the speed.

Similarly when an e. m. f. acts upon a circuit,† causing the current to increase, the increasing current reacts. The reacting e. m. f. is equal and opposite to the acting e. m. f., $L\frac{di}{dt}$ [equation (3)], which is causing the current to increase. This reacting e.

---

* In this equation L and i must be expressed in c. g. s. units because the unit of flux corresponding to the ampere-henry *is not much used.*

† Supposed to have zero resistance for the sake of simplicity of statement.

m. f. is called a self-induced e. m. f. The self-induced e. m. f. is therefore

$$e = -L\frac{di}{dt} \qquad (3)b$$

When a current is increasing $\left(\frac{di}{dt}\text{ positive}\right)$ the self-induced e. m. f. is opposed to the current, and when a current is decreasing $\left(\frac{di}{dt}\text{ negative}\right)$ the self-induced e. m. f. is in the direction of the current, exactly as in the case of a rotating wheel.

Proof of equation (5): If the current i is changing then from equation (5) we have $\frac{dN}{dt} = L\frac{di}{dt}$, but $-\frac{dN}{dt}$ is an e. m. f. e induced in the coil by the changing flux and, therefore, by the changing current. That is, $e = -L\frac{di}{dt}$, which, being identical to equation (3)b, shows that equation (5) is true.

11. Calculation of inductance in terms of magnetic flux per unit current.—According to equation (5) the inductance of a coil is equal to the quotient $\frac{N}{i}$ when N is the magnetic flux through the coil* due to the current i in the coil. There are important cases in which the flux through a coil due to a given current may be easily calculated and, therefore, the inductance of such a coil is easily determined.

Long Solenoid.—Consider a long cylindrical coil of wire of mean radius r, of length l and having z turns of wire per unit length. The field intensity in the coil is $f = 4\pi zi$ and the area of the opening of the coil is πr^2 so that the flux through the opening is $4\pi^2 r^2 zi (= N')$. The coil has lz turns, so that $N = lz.N' = 4\pi^2 r^2 z^2 li$; dividing this by i we have, according to equation (5),

$$L = 4\pi^2 r^2 z^2 l \qquad (6)$$

This equation is strictly true only for very long coils on which the wire is wound in a thin layer; the equation is, however, very

---

* That is, the flux through a mean turn multiplied by the number of turns of wire.

useful in enabling one to calculate easily the approximate inductance of even short thick coils.

Coil wound on an iron core.—A coil of Z turns of wire is wound on an iron ring l cm. in circumference (mean) and q cm.² in sectional area, as shown in Fig. 4. The coil produces through the ring a magnetic flux $N' = \dfrac{m.\,m.f.}{m.\,r.}$, where $m.\,m.\,f.$ $(= 4\pi Zi)$ is the magneto-motive force due to the coil, and $m.\,r.$ $\left(= \dfrac{l}{\mu}\cdot\dfrac{l}{q}\right)$ is the magnetic reluctance of the iron core, i being the current in the coil and μ the permeability of the iron. Therefore,

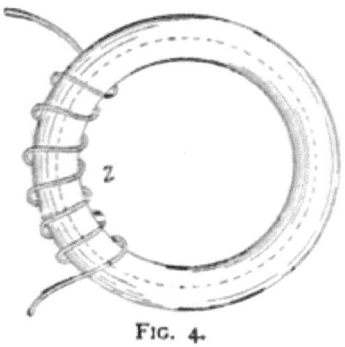

FIG. 4.

$$N = ZN' = \frac{4\pi\mu q Z^2 i}{l} = Li$$

or
$$L = \frac{4\pi\mu q Z^2}{l} \qquad (7)$$

Remark: The permeability μ of iron decreases with increasing magnetizing force. Therefore, the inductance of a coil wound on an iron core is not a definite constant as in case of a coil without an iron core.

12. Growth and decay of current in an inductive circuit.—When a torque is applied to a wheel the wheel gains speed until the whole of the applied torque is used to overcome the resistance of the air, etc. While the speed is increasing part of the applied torque overcomes this resistance and the remainder causes the speed to increase.

When an electromotive force is applied to a circuit the current in the circuit increases until the whole of the applied e. m. f. is used to overcome the resistance of the circuit. While the current is growing part of the applied e. m. f. overcomes resistance and the remainder causes the current to increase. Therefore,

$$E = Ri + L\frac{di}{dt} \qquad (8)$$

in which E is the applied e. m. f., i, is the instantaneous value of the growing current, R is the resistance of the circuit and L its inductance. Ri is the part of E used to overcome resistance and $L\frac{di}{dt}$ is the part of E used to make the current increase.

If a circuit of inductance L and resistance R with a given current is left to itself without any e. m. f. to maintain the current the current dies away or decays, and the e. m. f., Ri, which, at each instant overcomes the resistance, is the self-induced e. m. f. $-L\frac{di}{dt}$; so that at each instant $Ri = -L\frac{di}{dt}$ or

$$0 = Ri + L\frac{di}{dt} \qquad (9)$$

Examples: An e. m. f. of 110 volts acts on a coil of which the inductance is 0.04 henry and the resistance is 3 ohms. At the instant that the e. m. f. begins to act the actual current i in the coil is zero and the whole of the e. m. f. acts to increase the current, so that 110 volts = 0.04 henry $\times \frac{di}{dt}$ or $\frac{di}{dt}$ = 2750 amperes per second. When the growing current has reached a value of 30 amperes Ri is equal to 90 volts and the remainder of the 110 acts to cause the current to increase, that is, 20 volts = 0.04 henry $\times \frac{di}{dt}$ or $\frac{di}{dt}$ = 500 amperes per second.

If a current is established in this coil and the coil left to itself, short circuited, without any e. m. f. to maintain the current; then, as the decaying current reaches a value of, say, 30 amperes, the e. m. f. Ri is 90 volts and this e. m. f. is equal to $-L\frac{di}{dt}$ so that $\frac{di}{dt}$ is $-$ 2250 amperes per second.

13. Problem I.—An inductive circuit with a current flowing in it is left to itself, short circuited. At a certain instant, from which time is to be reckoned ($t = 0$), the value of the current is I. It is required to find an expression for the decaying current at each succeeding instant; the resistance R and the inductance L of the circuit being given.

Let i be the value of the current at the instant t. Then

$$i = I e^{-\frac{R}{L} \cdot t} \quad (10)$$

Proof: To establish the truth of equation (10) it is sufficient to show that $i = I$ when $t = 0$, and that equation (9) is satisfied. Substituting $t = 0$ in equation (10) we have $i = I$. Differentiating equation (10) we have $\frac{di}{dt} = -\frac{R}{L} I e^{-\frac{R}{L} \cdot t}$ or $\frac{di}{dt} = -\frac{R}{L} \cdot i$ or $Ri + L\frac{di}{dt} = 0$ which is equation (9).

The ordinates of the curve Fig. 5, show a decaying current.

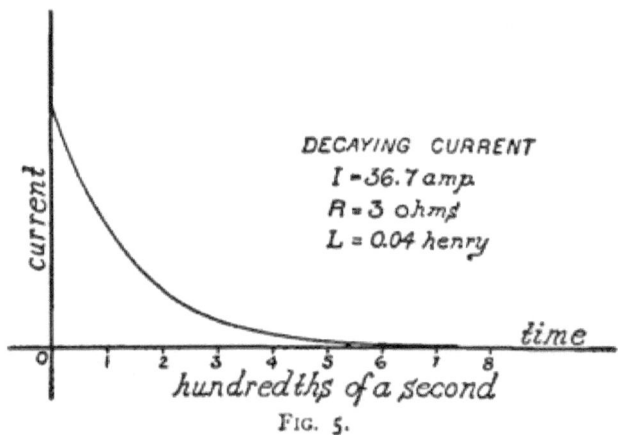

DECAYING CURRENT
I = 36.7 amp.
R = 3 ohms
L = 0.04 henry

FIG. 5.

14. Problem II.—A constant e. m. f. E is connected to a circuit of resistance R and inductance L. Required an expression for the growing current t seconds after the e. m. f. is connected to the circuit.

The required expression is

$$i = \frac{E}{R} - \frac{E}{R} e^{-\frac{R}{L} \cdot t} \quad (11)$$

Proof: To establish the truth of equation (11) it is sufficient to show that $i = 0$ when $t = 0$ and that equation (8) is satisfied.

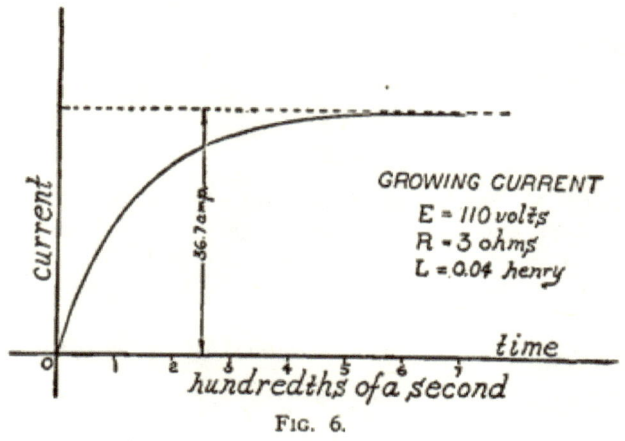

FIG. 6.

The ordinates of the curve, Fig. 6, show the values of a growing current.

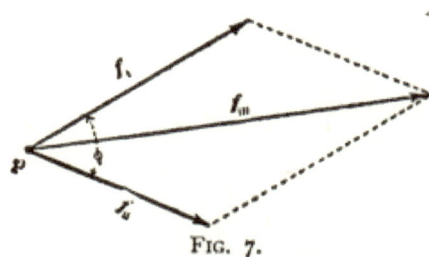

FIG. 7.

15. Energy of two coils. Mutual Inductance. Consider two separate coils in which currents i' and i'' respectively are flowing. The magnetic field f''', Fig. 7, at any given point p in the neighborhood is the resultant of the field intensities f' and f'' due to the respective currents so that from trigonometry

$$f'''^2 = f'^2 + f''^2 + 2f'f'' \cos \phi$$

Now the kinetic energy at p is proportional to f'''^2 or to $f'^2 + f''^2 + 2f'f'' \cos \phi$ so that the energy at the point consists of three parts proportional respectively to f'^2, to f''^2 and to $f'f''$. But the field intensities are proportional to the currents in the two coils so that the three parts of the energy at p are proportional respectively to i'^2 to i''^2 and to $i'\,i''$. The same is true of the energy at every other point of the magnetic field so that *the total kinetic energy of two coils is in three parts which are proportional respectively to i'^2, to i''^2 and to $i'\,i''$*, that is,

$$W = \tfrac{1}{2} L' i'^2 + \tfrac{1}{2} L'' i''^2 + M i' i'' \qquad (12)$$

in which W is the total kinetic energy of the two coils, i' and i'' the currents in the coils and $\tfrac{1}{2}L'$, $\tfrac{1}{2}L''$ and M are proportionality factors. The factors L' and L'' are the self inductances of the respective coils (previously defined) and the factor M is called the *mutual inductance* of the two coils.

From equation (12) it can be shown that

$$e'' = -M \frac{di'}{dt}$$

or

$$e' = -M \frac{di''}{dt}$$

(13)

in which e'' is the e. m. f. induced in one coil by a changing current $\frac{di'}{dt}$ in the other coil.

From equation (13) it can be shown that

$$N'' = Mi'$$

and

$$N' = Mi''$$

(14)

in which N'' is the flux through one coil (lines of force counted as many times as there are turns of wire) due to a current i' in the other coil.

Examples: The induction coil and the transformer depend for their action upon the mutual inductance of two coils, the *primary* and the *secondary* coils. In the theory and design of transformers it is not necessary to make explicit use of the idea of mutual inductance.

16. Electric charge.—The electric current in a wire is looked upon as a transfer of electric charge along the wire. The amount of electric charge Q which in t seconds passes a given point of a wire carrying a current i is

$$Q = it \qquad (15)$$

or the rate $\frac{dQ}{dt}$ at which the charge passes a given point on a wire is

$$\frac{dQ}{dt} = i. \qquad (16)$$

Units charge.—When i in equation (15) is expressed in amperes and t in seconds, Q is expressed in terms of a unit called the *coulomb*. That is, the coulomb is the amount of electric charge which passes in one second along a wire carrying one ampere. When i is expressed in c. g. s. units and t in seconds, Q is expressed in terms of the c. g. s. unit charge.

Measurement of electric charge.—An electric charge may be determined by measuring the current i which it will maintain during an observed time t. Then Q may be calculated from equation (15). The charge capacity of storage batteries is determined in this way. A very small charge cannot be measured by measuring the current i and the time t, for such a charge cannot maintain a steady measurable current for a sufficient time. A small electric charge is measured by allowing it to pass quickly through a galvanometer and observing the throw of the needle. The charge is sensibly proportional to the throw. A galvanometer used in this way is called a ballistic galvanometer.

17. Condensers. Electrostatic capacity.—When the terminals of a battery are connected to two metal plates, as shown in Fig. 8, a momentary current flows as indicated by the arrows and the electric charge which passes along the wire during this momentary current is stored upon the plates, for upon disconnecting the battery and connecting the plates with a wire a momentary reversed current may be observed. If a ballistic galvanometer be included in the circuit the amount of charge which passes into the plates may be measured. This amount of charge is proportional to the e. m. f. e, of the battery (other things being equal) that is

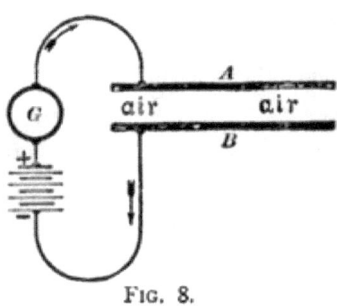

Fig. 8.

$$Q = Je \qquad (17)$$

in which Q is the electric charge which flows along the wire into the plates, e is the e. m. f. of the battery and J the proportionality factor. Two plates arranged in this way constitute what is called a *condenser* and the factor J is called the *electrostatic capacity* of the condenser. If, in equation (17) Q is expressed in coulumbs and e in volts, then J is expressed in terms of a unit called a *farad*. That is, a condenser has a capacity of one farad when one coulomb of electric charge is pushed into it by a battery of which the e. m. f. is one volt. The unit of capacity which is commonly used to express the capacities of condensers, electric cables, etc., is the microfarad. The microfarad is one millionth of a farad. The microfarad is used because the farad is too large a unit to use conveniently.

Condensers to have a large capacity (as much as a microfarad) are usually made up of alternate sheets of tinfoil and waxed paper or mica, as indicated in Fig. 9. Alternate metal sheets are connected together as shown, thus practically forming two plates of large area.

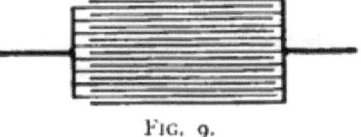

FIG. 9.

If capacities J_1, J_2, J_3, etc., are connected in *parallel* their combined capacity J is equal to $J_1 + J_2 + J_3 +$ etc. If capacities $J_1 J_2 J_3$, etc., are connected in *series* their combined capacity J is obtained from the expression $\frac{1}{J} = \frac{1}{J_1} + \frac{1}{J_2} + \frac{1}{J_3} +$ etc. It will be noted that capacities combine in a manner just the opposite of that of resistances.

18. Mechanical and electrical analogies.—The analogy between moment of inertia and inductance as pointed out in the discussion of inductance is but a small part of an extended analogy between pure mechanics and electricity. This extended analogy is here briefly outlined.

$x = vt$ (1)	$\phi = \omega t$ (2)	$q = it$ (3)
in which x is the distance traveled in t seconds by a body moving at velocity v.	in which ϕ is the angle turned in t seconds by a body turning at angular velocity ω.	in which q is the electric charge which in t seconds flows through a circuit carrying a current i.
$W = Fx$ (4)	$W = T\phi$ (5)	$W = Eq$ (6)
in which W is the work done by a force F in pulling a body through the distance x.	in which W is the work done by a torque T in turning a body through the angle ϕ.	in which W is the work done by an e.m.f. E in pushing a charge q through a circuit.
$P = Fv$ (7)	$P = T\omega$ (8)	$P = Ei$ (9)
in which P is the power developed by a force F acting upon a body moving at velocity v.	in which P is the power developed by a torque T acting on a body turning at angular velocity ω.	in which P is the power developed by an e.m.f. E in pushing a current i through a circuit.
$W = \frac{1}{2} mv^2$ (10)	$W = \frac{1}{2} K\omega^2$ (11)	$W = \frac{1}{2} Li^2$ (12)
in which W is the kinetic energy of a mass m moving at vlocity ev.	in which W is the kinetic energy of a wheel of moment of inertia K turning at angular velocity ω.	in which W is the kinetic energy of a coil of inductance L carrying a current i.

$$F = m\frac{dv}{dt} \quad (13)$$

in which F is the force required to cause the velocity of a body of mass m to increase at the rate $\frac{dv}{dt}$

$$T = K\frac{d\omega}{dt} \quad (14)$$

in which T is the torque required to cause the angular velocity of a wheel of moment of inertia K to increase at the rate $\frac{d\omega}{dt}$

$$E = L\frac{di}{dt} \quad (15)$$

in which E is the e. m. f. required to cause a current in a coil of inductance L to increase at the rate $\frac{di}{dt}$

$$x = aF \quad (16)$$

$$\frac{4\pi^2 m}{\tau^2} = \frac{1}{a} \quad (19)$$

$$\varphi = bT \quad (17)$$

$$\frac{4\pi^2 K}{\tau^2} = \frac{1}{b} \quad (20)$$

$$q = JE \quad (18)$$

$$\frac{4\pi^2 L}{\tau^2} = \frac{1}{J} \quad (21)$$

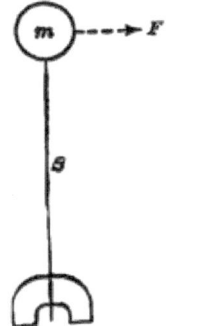

Fig. *a*. Art. 18.

A body of mass m is supported by a flat spring S clamped in a vise as shown in Fig. *a*. A force F pushing sidewise on m moves it a distance x which is proportional to F according to equation (16). When started the body m will continue to vibrate back and forth and the period τ of its vibrations is determined by equation (19).

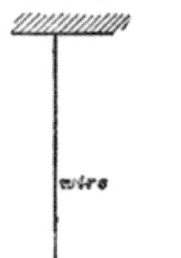

Fig. *b*. Art. 18.

A body of moment of inertia K is hung by a wire as shown in Fig. *b*. A torque T acting on the body will turn the body and twist the wire through an angle ϕ which is proportional to T according to equation (17). When started, the body will vibrate about the wire as an axis and the period τ of its vibrations is determined by equation (20).

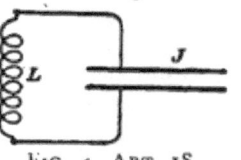

Fig. *c*. Art. 18.

A condenser J is connected to the terminals of a coil of inductance L as shown in Fig *c*. An e. m. f. E acting anywhere in the circuit pushes into the condenser a charge q which is proportional to E according to equation (18). When started the electric charge will surge back and forth through the coil constituting what is called an oscillatory current and the period of one oscillation is determined by equation (21).

Problems.

1. The intensity of the magnetic field in the air gap between the pole face and the armature core of a dynamo is 5,000 c. g. s. units and the pole face is 10 cm. × 20 cm. Required the magnetic flux from pole face to armature core.

2. A coil of an alternator armature has 20 turns and engages the whole 1,500,000 lines which flow from a pole of the field magnet. In $\frac{1}{250}$ second this coil moves from a north pole to an adjacent south pole of the field magnet when the flux is reversed. Calculate the average e. m. f. in the coil during this interval.

3. Find the approximate inductance, in henrys, of a cylindrical coil 25 cm. long, 5 cm. mean diameter wound with one layer of wire containing 150 turns.

4. Calculate the kinetic energy in joules of a current of 20 amperes in the above coil.

5. The above coil is connected to 110 volt mains, find the rate, in amperes per second, at which the current begins to increase in the coil.

6. Calculate the rate at which the current is increasing (problem 5) when it has reached the value of 10 amperes, the resistance of the coil being 0.25 ohm.

7. A coil of which the resistance is 2.5 ohms and the inductance 0.04 henry has a current started in it. The coil is then short circuited and the current left to die away. Calculate the rate, in amperes per second, at which the current is decreasing as it passes the value of 10 amperes.

8. A coil of wire has an inductance of .035 henry, calculate the magnetic flux through the coil due to a current of 5 c. g. s. units in the coil. If the coil has 1,500 turns of wire calculate number of lines of flux through a mean turn.

9. A condenser has a capacity of 1.2 microfarads, calculate the charge which is pushed into this condenser by an e. m. f. of 1,000 volts, and calculate the time during which this would maintain a current of one ampere.

10. The field coils of a shunt dynamo have a resistance of 100 ohms and an inductance of 20 henrys. An e. m. f. of 500 volts is applied. Calculate the time required for the current to reach a value of 4 amperes.

CHAPTER II.

THE SIMPLE ALTERNATOR.

19. The alternator consists essentially of a magnet, near the poles of which a coil of wire is moved in such a way that the magnetic flux from the poles passes through the coil, first in one direction and then in the other. This varying magnetic flux induces an electromotive force in the coil, first in one direction and then in the other. This e. m. f., called an *alternating* e. m. f., produces an *alternating* current in the coil and in the circuit, which is connected to the terminals of the coil.

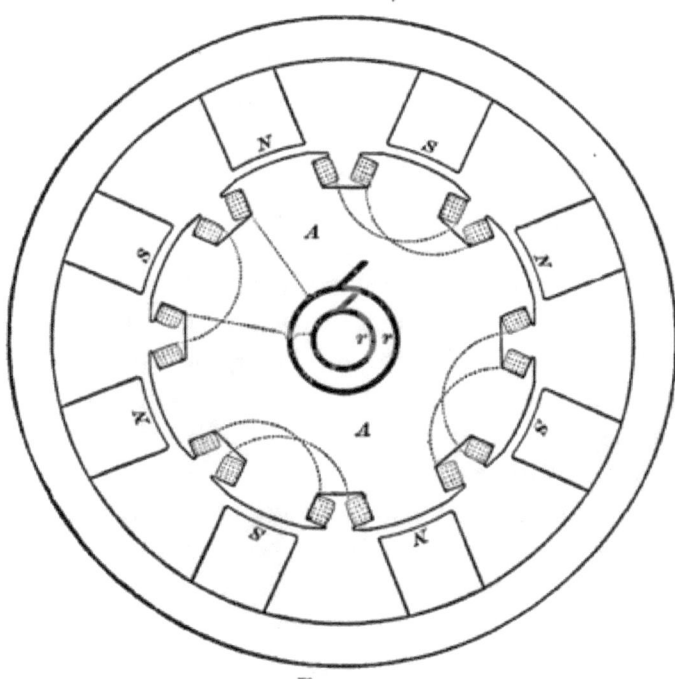

FIG. 10.

A common type of alternator consists of a multipolar electromagnet (the *field magnet*) of which the poles project radially in-

(18)

ward towards the passing teeth of a rotating laminated iron drum A (the *armature*), as shown in Fig. 10. On the armature shaft, at one end of the armature, are mounted two insulated metal rings $r\,r$ (*collecting rings*), upon which metal springs (*brushes*) rub, keeping continuous contact with the terminals of an external circuit. The ends of the armature wire are fastened to the respective collecting rings, the armature coils being wound around the teeth, as shown.

The e. m. f.'s induced in adjacent coils are in opposite directions and the coils are so connected together that these e. m. f.'s do not oppose each other. This is done by reversing the connections of every alternate coil, as indicated by the dotted lines connecting the coils.

The field magnet of the alternator is excited by a continuous current from some independent source, generally an auxiliary dynamo called the *exciter*. The type of armature winding shown in Fig. 10 is known as the *concentrated* winding. In this type of winding the armature conductors are grouped in a few heavy coils. Alternator armatures are also made in which the winding is distributed in a large number of small slots. This type of winding is known as the *distributed* winding; it is described in Chapter IX.

20. Speed and frequency.—The e. m. f. of an alternator passes through a set of positive values, while a given coil of the armature is passing from a south to a north pole of the field magnet, and through a *similar* set of negative values, while the coil is passing from a north pole to a south pole, or *vice versa*. The complete set of values, including positive and negative values, through which an alternating e. m. f. (or alternating current) repeatedly passes, is called a *cycle*. The number of cycles per second is called the *frequency f*.

Let p be the number of *pairs* of field magnet poles, n the revolutions per second of the armature, and f the frequency of the e. m. f. of the alternator. Then

$$f = pn \qquad (18)$$

This is evident when we consider that the e. m. f. passes through a complete cycle of values while an armature tooth is passing from a north pole to the next north pole, so there are p cycles of e. m. f. for each revolution of the armature. The frequency depends only on the speed and number of poles and is not dependent in any way upon the style of armature or armature winding.

21. Electromotive force and current curves.—The successive instantaneous values of the e. m. f. of an alternator may be reppresented by the ordinates of points on a curve, the abscissas representing time elapsed from some chosen epoch; the resulting curve is called the *e. m. f. curve* of the alternator. In a similar manner the successive instantaneous values of an alternating current may be represented by ordinates and the elapsed time by abscissas giving a *current curve*. These curves are determined with the help of the *contact maker* as explained in Art. 25.

Examples: The full line curve, Fig. 11, represents the e. m. f.

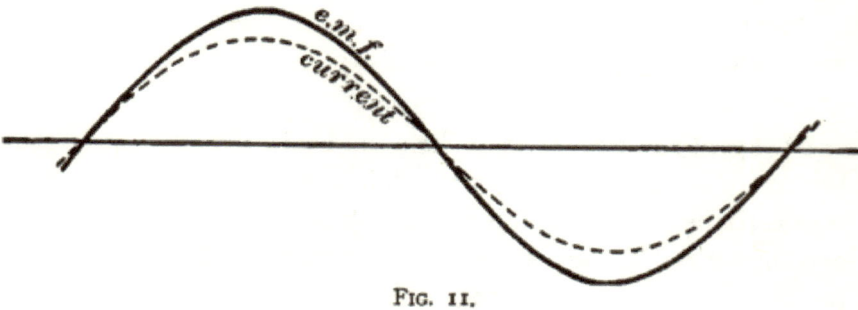

FIG. 11.

of a smooth core alternator and the dotted curve represents the current which this e. m. f. produces in a non-inductive circuit; this current is at each instant equal to the e. m. f. divided by the resistance of the circuit so that the current is a maximum when the e. m. f. is a maximum, the current is said to be *in phase* with the e. m. f. as is explained in Chapter IV.

The full line curve, Fig. 12, represents the e. m. f. of a smooth core armature and the dotted curve represents the current which this e. m. f. produces in an inductive circuit. In this case part

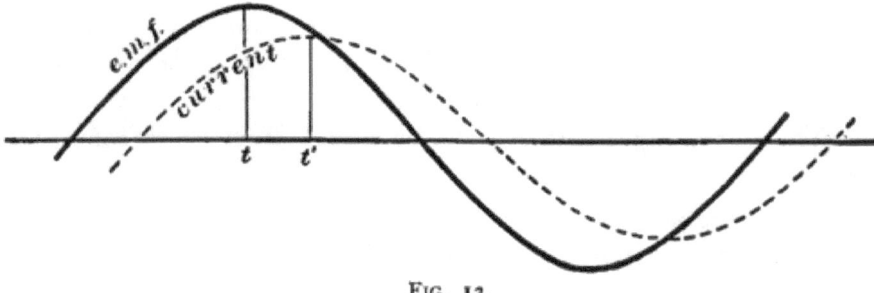

FIG. 12.

of the e. m. f. is, at each instant, used to cause the current to increase or decrease. The part so used is $L\dfrac{di}{dt}$ according to equation (3), and the remainder, equal to Ri, is used to overcome

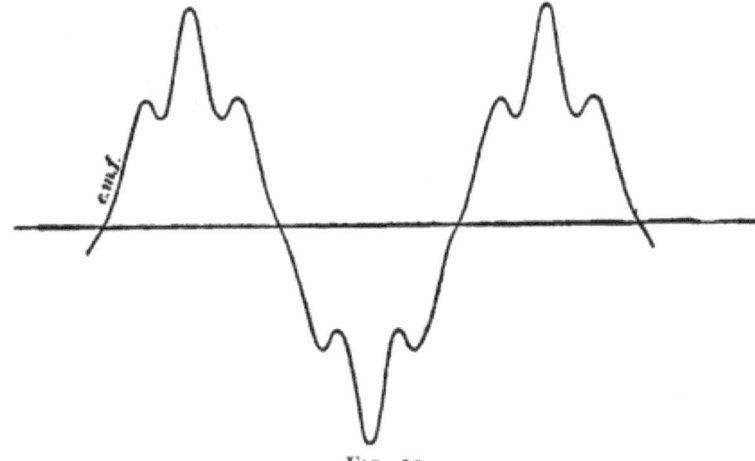

FIG. 13.

the resistance of the circuit. When the current is zero then all the e. m. f. is used to cause the current to change since Ri is zero. When $\dfrac{di}{dt}$ is zero, the current is at its maximum or minimum value, and, at this instant, all the e. m. f. is used to overcome the resistance of the circuit since $L\dfrac{di}{dt}$ is zero. It is to be particularly noted that the time t', Fig. 12, at which the current reaches its maximum value is later than the time t at which the e. m. f. reaches its maximum value. In some cases, however, the current may reach its maximum value before the e. m. f.

The curve, Fig. 13, represents the e. m. f. of an alternator with a toothed armature core.

22. Instantaneous and average power expended in an alternating current circuit.—Let e be the value at a given instant of an alternating e. m. f. and i the value of the current at the same instant. Then ei is the power in watts which, at the given instant, is being expended in the circuit, and the average value of ei is the average power expended in the circuit. In Fig. 14 the full

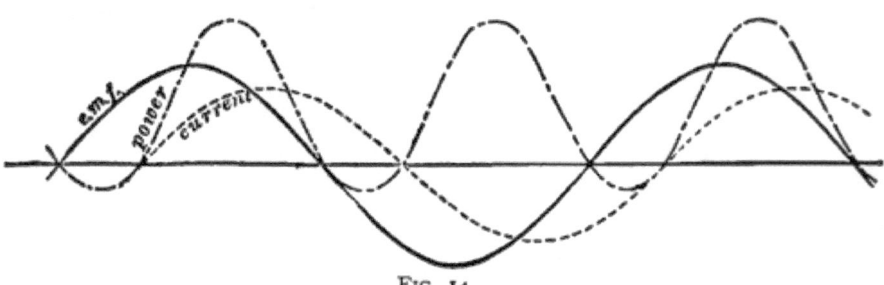

FIG. 14.

line curve represents the e. m. f. of an alternator and the dotted curve represents the current produced by the alternator in a circuit having inductance. The ordinates of the dot dash curve represents the successive instantaneous values of the power ei. As is shown in the figure, the power has both positive and negative values, the alternator does work on the circuit when ei is positive and the circuit returns power to the alternator when ei is negative, and, of course, while ei is negative the dynamo is momentarily a motor and may for the moment return power to the fly-wheel of the engine.

When the inductance of the circuit of an alternator is very large the e. m. f. and current curves are as shown in Fig. 15, the instantaneous power *ei* passes through approximately similar sets of positive and negative values as shown by the dot and dash curves, and the average power is zero.

23. Average values and effective values.—The average value of an alternating current or e. m. f. is zero, inasmuch as similar

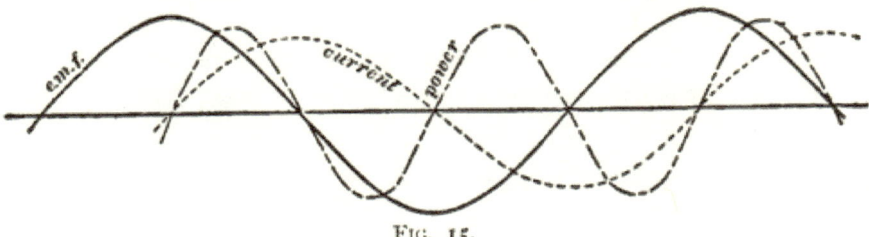

FIG. 15.

sets of positive and negative values occur. The average value of an e. m. f. or current *during the positive (or negative) part of a cycle* is usually spoken of briefly as the average or mean value and is not zero.

Effective values.—Consider an alternating current of which the instantaneous value is *i*. The rate at which heat is generated in a circuit through which the current flows is Ri^2, where R is the resistance of the circuit, and the average rate at which heat is generated in the circuit is R multiplied by the average value of i^2. A continuous current which would produce the same heating effect would be one of which the square is equal to the average value of i^2 or of which the actual value is equal to $\sqrt{\text{average } i^2}$. This square root of the average square of an alternating current is called the *effective* value of the alternating current. Similarly the square root of the average square of an alternating e. m. f. is called the *effective value* of the alternating e. m. f.

Ammeters and voltmeters used for measuring alternating currents and alternating e. m. f.'s always give effective values as is shown in Chapter III; and in specifying an alternating e. m. f. or current its effective value is always used.

Remark.—The ratio *effective value* divided by *average value* of an alternating e. m. f., or current, is in most cases approximately equal to $\dfrac{\pi}{2\sqrt{2}} = 1.11$ as is shown in Chapter IV.

Example.—Consider the successive instantaneous values (separated by equal time intervals) of an alternating current. The sum of these values divided by their number gives their average value. Square each instantaneous value. Add these squares, divide by their number and extract the square root, and the result is the square-root-of-average-square, or effective, value of the current.

24. The fundamental equation of the alternator.—The equation which expresses the effective value of the e. m. f. of an alternator in terms of the armature speed n, the number of pairs of field magnet poles p, the flux N from one pole of the field magnet and the total number of armature conductors C which cross the face of the armature is called the fundamental equation of the alternator. This equation is important in designing. It is derived as follows for the case in which the armature conductors are concentrated in $2p$ slots, one to each field pole as shown in Fig. 16.

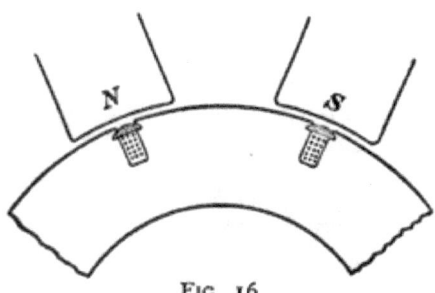

Fig. 16.

Let N be the lines of flux from one pole, then one armature conductor in one revolution, cuts $2p.N$ lines,* and in one second it cuts $2pNn$ lines, which is the *average* e. m. f. (in c. g. s. units) induced in one armature conductor. We have, therefore,

---

* Since we are concerned with the average value during half a cycle the change of sign during the two halves of a cycle is to be ignored and the flux from north and from south poles is to be treated without regard to sign

THE SIMPLE ALTERNATOR.

$$\text{Average* e. m. f. of alternator} = \frac{2pNCn}{10^8} \quad (19)$$

The ratio, effective e. m. f. divided by average e. m. f., is for commercial alternators, approximately equal to 1.11.† Therefore, the effective e. m. f. of an alternator with concentrated armature winding is approximately

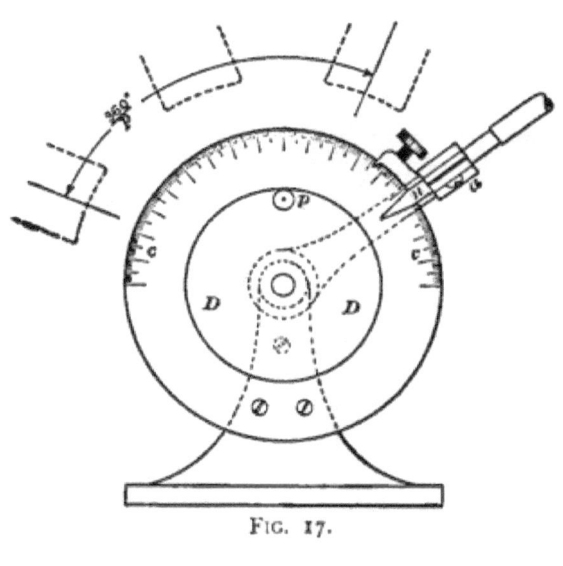

FIG. 17.

$$E = \frac{2.22pNCn}{10^8} \quad (20)$$

or, since pn is the frequency according to equation (18) we have

$$E = \frac{2.22NCf}{10^8} \quad (21)$$

In which N is the magnetic flux from *one* pole, and C is the total number of conductors on the armature which are connected in series. Sometimes it is more convenient to have the equation

---

* That is the average during half a cycle as explained in Art. 22. The average during a whole cycle is zero.

† See Arts. 23, 46 and 47.

given in terms of *armature turns* instead of *armature conductors*. The formula then becomes

$$E = \frac{4.44 NTf}{10^8} \qquad (22)$$

T being the number of armature turns in series between the collector rings.

25. Experimental determination of alternator e. m. f. curves. *The contact maker.*—A disc DD, Figs. 17 and 18, fixed to and rotating with the armature shaft, carries a pin p which makes momentary electrical contact, once per revolution, with a jet of conducting liquid which issues from a nozzle n. This nozzle is carried on a pivoted arm a, and can be moved at will, its position

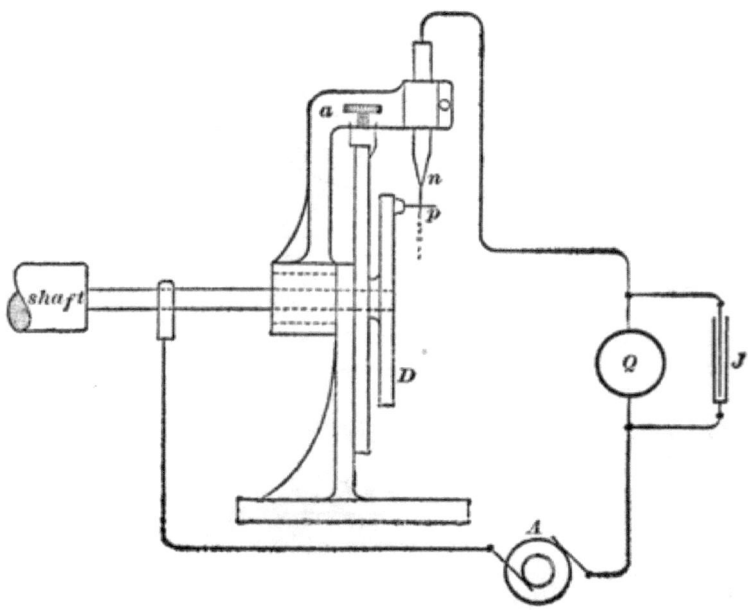

FIG. 18.

being read off the divided circle cc. One terminal of an electrostatic voltmeter Q is connected directly to one brush of the alternator, while the other terminal of the voltmeter is connected through the jet and pin to the other brush of the alternator as shown in Fig. 18. The voltmeter then indicates the value of the e. m. f. of the alternator at the instant of contact of jet and pin. By shifting the jet, step by step, around the circle successive instantaneous values of the e. m. f. may be determined. The e. m. f. passes through a complete cycle of values while the jet is shifted $\frac{1}{p}$ of a revolution, p being the number of pairs of poles of the alternator. In order that the e. m. f. acting upon the electrostatic voltmeter may not fall off appreciably in the intervals between successive contacts of pin and jet, a condenser J is connected as shown in Fig. 18. The indications of an electrostatic voltmeter are not accurate for small deflections and in using such an instrument for measuring a comparatively small e. m. f. a battery of known e. m. f. may be connected in the circuit so as to raise the e. m. f. to an accurately measurable value.

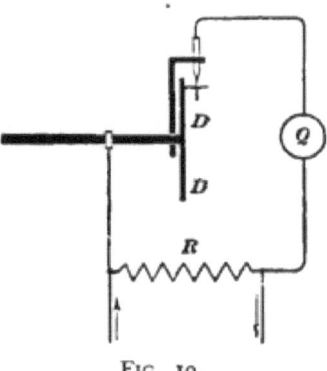

FIG. 19.

In the determination of an alternating current curve, the current is sent through a non-inductive resistance R, Fig. 19, and the e. m. f. between the terminals of this resistance is determined as before, the disc DD being fixed to the armature shaft of the alternator which is furnishing the current. The current at each instant is equal to the e. m. f. divided by R.

PROBLEMS.

1. An alternator has 16 poles and its speed is 900 revolutions per minute. What is the frequency of its e. m. f. ?

2. An alternator has 8 poles and its speed is 900 revolutions per minute. The flux from one pole is 2,200,000 lines. The armature has 1000 conductors (wound in 8 slots) all of which

are connected in series. What will be the effective e. m. f. obtained between the collector rings?

3. An alternator has 10 poles and runs at a speed of 1500 revolutions, generating 2000 volts. The flux from one pole is 2,250,000 lines. How many turns must there be on the armature if they are all connected in series?

4. The following are the instantaneous values of an e. m. f. taken at equal intervals during half a cycle: 0, 30, 60, 80, 90, 100, 90, 80, 60, 30, 0 volts. The corresponding values of the current are -45, -25, 0, 25, 50, 65, 75, 75, 70, 60, 45 amperes.

Find the effective value of the e. m. f. Find the instantaneous values of the power and find the average power.

CHAPTER III.

ALTERNATING AMMETERS, VOLTMETERS AND WATTMETERS.

26. The hot wire ammeter and voltmeter.*—In these instruments the current to be measured is sent through a stretched wire. This wire, heated by the current, lengthens and actuates a pointer which plays over a divided scale.

The hot wire instrument, when calibrated by continuous currents, indicates effective values of alternating currents; and when calibrated by continuous e. m. f.'s it indicates effective values of alternating e. m. f.'s.

Proof: Consider an alternating current and a continuous current C which give the same reading. *These currents generate heat in the wire at the same average rate.* This rate is RC^2 for the continuous current and $R \times$ average i^2 for the alternating current, i being the instantaneous value of the alternating current. Therefore $RC^2 = R \times$ average i^2 or $C^2 =$ average i^2 or $C = \sqrt{\text{average } i^2}$. Q. E. D.

The proof for e. m. f.'s is similar to this proof for currents.

Remark: The only hot wire instrument which is much used is the *Cardew voltmeter*. Such instruments need to be frequently re-calibrated, and are, therefore, not very satisfactory.

27. The electro-dynamometer used as an ammeter.—The essential parts of the electro-dynamometer are shown in Figs. 20 and 21. These figures show the arrangement of the parts in Siemens' type of instrument. The coil A is held stationary by the frame of the instrument while the coil B is mounted at right angles to A and is hung from a suspension. This movable coil is provided with flexible or mercury cup connections $a\,a$ and the current to be measured is sent through both coils in series. The force

---

* All voltmeters, except the electrostatic voltmeter, are essentially ammeters. That is, the e. m. f. to be measured produces a current which actuates the instrument. The scale over which the pointer plays may be arranged to indicate either the value of the current or the value of the e. m. f.

action between the coils is balanced by carefully twisting a helical spring *b*, one end of which is attached to the coil *B* and the other to the torsion head *c*. The observed angle of twist necessary to bring the swinging coil to its zero position is read off by means of the pointer *d* and the graduated scale *e*. The pointer *f* attached

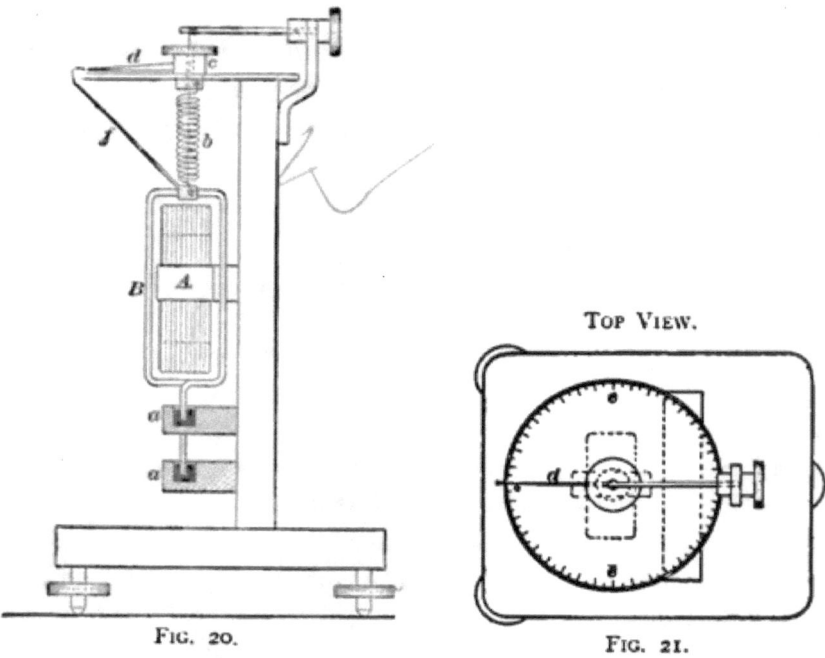

FIG. 20.   FIG. 21.

to the coil shows when it has been brought to its zero position. The observed angle of twist of the helical spring affords a measure of the force action between the coils, and the current is proportional to the square root of this angle of twist. In other forms of electro-dynamometer the force action between the coils moves the suspended coil and causes the attached pointer to play over a divided scale.

The electro-dynamometer when standardized by direct currents indicates effective values of alternating currents.

Proof: A given deflection of the suspended coil depends upon a definite average or constant force action between the coils. The force action due to a constant current

c is kc^2 (proportional to c^2) and the average force action due to an alternating current is $k \times$ average i^2, so that if these currents give the same deflection we have $kc^2 = k \times$ average i^2, or $c^2 =$ average i^2, or $c = \sqrt{\text{average } i^2}$. Q. E. D.

Remark: The electro-dynamometer is the standard instrument for measuring alternating currents and it is always used in refined measurements. The most satisfactory type of electro-dynamometer is Kelvin's Balance.

28. The electro-dynamometer used as a voltmeter.—When used as a voltmeter the coils of the electro-dynamometer are made of fine wire, and an auxiliary non-inductive resistance is usually connected in series with the coils.

When the inductance of the electro-dynamometer coils is small such an instrument, when calibrated by continuous e. m. f.'s, indicates effective values of alternating e. m. f.'s.

When it is certain that the inductance of an electro dynamometer is negligibly small the instrument may be used in refined alternating e. m. f. measurements.

Inductance error of the electro-dynamometer used as a voltmeter.—An electro-dynamometer which has been calibrated by continuous electromotive force indicates less than the effective value of an alternating e. m. f. The following discussion of this error for the case of harmonic e. m. f. presupposes a knowledge of Chapters IV and V. Let $\mathbf{E}$ be the reading of an electro-dynamometer voltmeter when an alternating e. m. f. (harmonic) of which the effective value is E is connected to its terminals. That is, $\mathbf{E}$ is the continuous e. m. f. which gives the same deflection as E and since E gives the same deflection as $\mathbf{E}$ it follows that the effective current produced by E is equal to the continuous current produced by $\mathbf{E}$; that is,

$$\frac{\mathbf{E}}{R} = \frac{E}{\sqrt{R^2 + \omega^2 L^2}} \qquad (1)$$

in which R is the total resistance of the instrument, L its inductance, and $\omega = 2\pi f$ where f is the frequency of the alternating e. m. f. Solving equation (1) for E we have

$$E = \frac{\sqrt{R^2 + \omega^2 L^2}}{R} \cdot \mathbf{E} \qquad (23)$$

That is, the reading of the instrument must be multiplied by the factor

$$\frac{\sqrt{R^2 + \omega^2 L^2}}{R}$$

to give the true effective value of an harmonic alternating e. m. f.

Plunger type voltmeters have inductance errors also.

29. The electrostatic voltmeter.—Two insulated metal plates which are connected to the terminals of a battery, or to any source of e. m. f., attract each other with a force which is strictly proportional to the square of the e. m. f. This principle is applied in the *electrostatic voltmeter*, which consists essentially of a fixed plate and a suspended plate to which a pointer is attached. The terminals of the e. m. f. to be measured are connected to these plates.

Such an instrument, when calibrated by continuous e. m. f., indicates effective values of alternating e. m. f.

Proof: A given deflection of the suspended plate depends upon a definite average or constant force action between the plates. The force action due to a constant e. m. f. E is KE^2 (proportional to E^2) and the average force action due to an alternating e. m. f. e, is $K \times$ average e^2. If these e. m. f.'s give equal deflections the force KE^2 is equal to the average force $K \times$ average e^2 so that $E^2 =$ average e^2, or $E = \sqrt{\text{average } e^2}$. Q. E. D.

The electrostatic voltmeter is the standard instrument for measuring alternating e. m. f.'s, especially for the measurement of

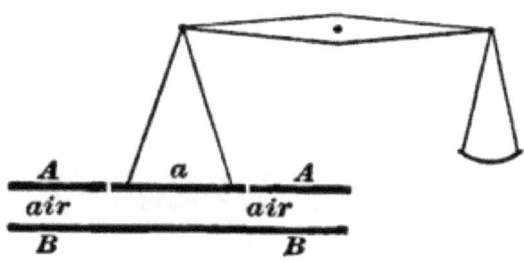

Fig. 22.

very high e. m. f. Further, with high e. m. f.'s the electrostatic attraction of parallel metal plates is great enough to be accurately measured by a balance and in this case the e. m. f. between the plates (constant e. m. f., or effective value of an alternating e. m. f.) may be calculated independently of calibrations of any kind. An instrument arranged for the absolute* measurement of e. m. f. in this way is called an *absolute electrometer*.

---

* That is, the measurement in terms of mechanical units of force, distance, etc.

The absolute electrometer consists of two parallel metal plates, AaA and BB, Fig. 22. The central portion a of the upper plate, while remaining in electrical communication with AA, is detached and suspended from one arm of a balance beam as shown. The e. m. f. E between AaA and BB is given by the formula

$$E^2 = \frac{Fd^2}{2400\pi a} \qquad (24)$$

in which F is the observed downward pull on a in dynes, d is the distance apart of the plates in centimeters and a is the area in cm^2 of the detached portion a.

30. The spark gauge.—The high e. m. f.'s used in break-down tests are usually measured by means of the *spark gauge*. This consists of an adjustable air gap which is adjusted until the e. m. f. to be measured is just able to strike across in the form of a spark. The e. m. f. is then taken from empirical tables based upon previous measurements of the e. m. f. required to strike across various widths of gap. In the spark gauge of the General Electric Co. the spark gap is between metal points, one of which is attached to a micrometer screw by means of which the gap space may be adjusted and measured. The striking distance in any spark gauge varies greatly with the condition of the points. It is, therefore, necessary to see that the points are well polished before taking measurements.

31. Plunger type ammeters and voltmeters.—In instruments of this type the current to be measured passes through a coil of wire which magnetizes and attracts a movable piece of soft iron to which the pointer is fixed.

A plunger meter (ammeter or voltmeter) should be calibrated under the conditions in which it is to be used. Thus, if a plunger instrument is to be used as an ammeter for alternating currents of a given frequency it should be calibrated by currents of this frequency, these currents being, for the purpose of the calibration, measured by a standard alternating current ammeter such as an electrodynamometer. The indications of a plunger instrument do not, however, vary greatly with frequency and such instruments are used for approximate measurements without regard to frequency.

The Thomson inclined coil meter of the General Electric Co. is of the plunger type. The essential parts of this instrument are shown in Fig. 23. A coil A, through which flows the current

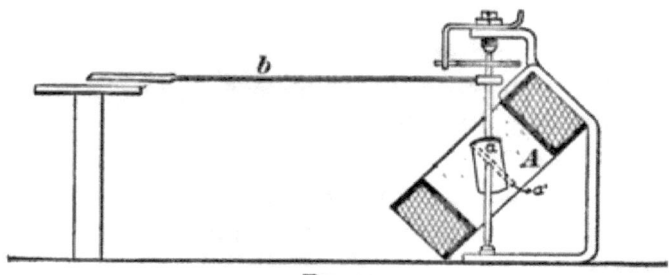

FIG. 23.

to be measured, is mounted with its axis inclined as shown. A vertical spindle mounted in jeweled bearings and controlled by a hair-spring passes through the coil, and to this spindle are fixed a pointer b and a vane of thin sheet-iron a. This vane of iron is mounted obliquely to the spindle. When the pointer is at the zero point of the scale the iron vane a lies nearly across the axis of the coil, and when a current passes through the coil the vane tends to turn until it is parallel to the axis of the coil, thus turning the spindle and moving the attached pointer over the calibrated scale.

32. The potential method for measuring alternating current.—The alternating current to be measured is passed through a known non-inductive resistance R and the e. m. f. between the terminals of this resistance is measured by a voltmeter. The current (effective value) is then equal to the e. m. f. (effective value) divided by the resistance.

33. The calorimetric method for measuring alternating current.—The current to be measured is passed through a known resistance which is submerged in a calorimeter by means of which the heat H which is generated in the resistance in an observed interval of time t is determined. This heat being expressed in joules we have

$$H = I^2 Rt \qquad (25)$$

in which I is the effective value of the current.

MEASUREMENTS OF POWER IN ALTERNATING CIRCUITS.*

34. The three-voltmeter method.—A non-inductive resistance R, Fig. 24, is connected in series with the circuit bc in which the power P, to be determined, is expended. The e. m. f.'s E_1 between ab, E_2 between bc, and E_3 between ac, are observed by means of a voltmeter as nearly simultaneously as possible. Then

$$P = \frac{E_3^2 - E_1^2 - E_2^2}{2R} \qquad (26)$$

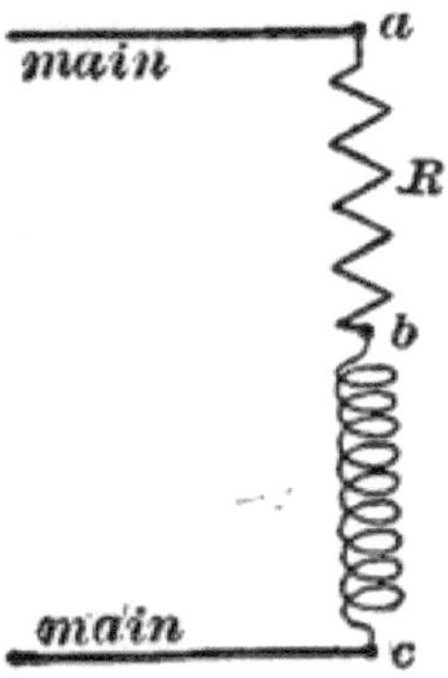

Proof: Let e_1, e_2 and e_3 be the instantaneous e. m. f.'s between ab, bc and ac respectively, then

$$e_3 = e_1 + e_2 \qquad \text{(i)}$$

or $\qquad e_3^2 = e_1^2 + 2e_1 e_2 + e_2^2 \qquad \text{(ii)}$

or $\qquad$ Average $e_3^2 =$ average $e_1^2 + 2$ average $e_1 e_2 +$ average e_2^2† $\qquad$ (iii)

but $E_3^2 =$ average e_3^2, $E_1^2 =$ average e_1^2, and $E_2^2 =$ average e_2^2. Further $\frac{e_1}{R}$ is the instantaneous current in abc, $\frac{e_1}{R} \cdot e_2$ is the instantaneous power expended in bc and average $\left(\frac{e_1}{R} \cdot e_2 \right)$ or $\frac{1}{R} \times$ average $(e_1 e_2)$ is the average power P expended in bc so that average $(e_1 e_2) = RP$. Therefore equation (iii) becomes

$$E_3^2 = E_1^2 + 2RP + E_2^2 \qquad \text{(iv)}$$

or $\qquad P = \dfrac{E_3^2 - E_1^2 - E_2^2}{2R} \qquad$ Q. E. D.

FIG. 24.

---

* In alternating circuits power cannot be measured by means of an ammeter and a voltmeter as in the case of direct current for the reason that the power expended is in general less than the product of effective e. m. f. into effective current on account of the difference in phase of the current and e. m. f.

† For proof of (iii) see proposition Art. 45.

35. The three-ammeter method for measuring power.—The circuit CC, Fig. 25, in which the power P, to be measured is expended, is connected in parallel with a non-inductive resistance R and three ammeters are placed as shown. Then

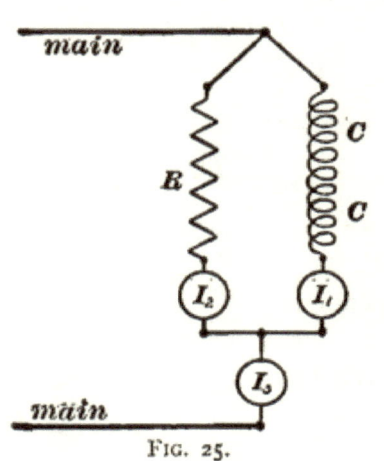

Fig. 25.

$$P = \frac{R}{2}(I_3^2 - I_1^2 - I_2^2) \quad (27)$$

in which I_1, I_2 and I_3 are the currents indicated by the three ammeters.

Proof: Let i_1, i_2 and i_3 be the instantaneous values of the currents I_1, I_2 and I_3. Then

$$i_3 = i_1 + i_2 \quad (i)$$

or

$$i_3^2 = i_1^2 + 2i_1i_2 + i_2^2 \quad (ii)$$

or average i_3^2 = average i_1^2 + 2 average $(i_1 i_2)$ + average i_2^2.

But $I_1^2 =$ average i_1^2, $I_2^2 =$ average i_2^2, and $I_3^2 =$ average i_3^2.

Further, the instantaneous e. m. f. between the terminals of R or of CC is Ri_2 so that Ri_2i_1 is the instantaneous power expended in CC and $R \times$ average $(i_1 i_2)$ is the average power P, expended in CC. Therefore, average $(i_1, i_2) = \frac{P}{R}$ and equation (iii) becomes:

$$I_3^2 = I_1^2 + \frac{2P}{R} + I_2^2 \quad (iv)$$

or

$$P = \frac{R}{2}(I_3^2 - I_1^2 - I_2^2) \quad \text{Q. E. D.}$$

Combination method.—The three-ammeter method for measuring power may be modified by using the potential method for measuring I_2, Fig. 25. In this case the e. m. f. between the terminals of R, Fig. 25, is measured by means of a voltmeter so that $I_2 = \frac{E_2}{R}$ where E_2 is the voltmeter reading.

36. The Wattmeter.—The Wattmeter is an electrodynamometer of which one coil a, Fig. 26, made of fine wire, is connected to the terminals of the circuit CC in which the power to be measured is expended. The other coil b made of large wire is connected in series with CC as shown. The fine wire coil a is movable and carries the pointer which indicates the watts expended in CC.

Such an instrument when calibrated with continuous current and e. m. f. indicates power accurately when used with alternating currents, provided the inductance of the coil a together with the auxiliary resistance r is small.*

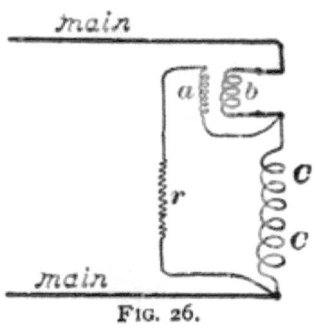

Fig. 26.

Proof.—A given deflection of the movable coil a depends upon a certain average or constant force action between the coils. Consider a continuous e. m. f. E which produces a current $\frac{E}{r}$ in a and a current C in CC and b. The force action between the coils is proportional to the product of the currents in a and b, that is, the force action is $k \cdot \frac{E}{r} \cdot C$, where k is a constant.

Consider an alternating e. m. f. of which the instantaneous value is e; this produces a current $\frac{e}{r}$ through a (provided the inductance of a is zero) and a current i in CC and b. The instantaneous force action between the coils is $k \cdot \frac{e}{r} \cdot i$ and the average force action is $\frac{k}{r} \cdot$ average (ei). If this alternating e. m. f. gives the same deflection as the continuous e. m. f. then

$$\frac{k}{r} \times \text{average } (ei) = EC\frac{k}{r}$$

or average $(ei) = EC$.

That is the given deflection indicates the same power whether the currents are alternating or direct. Q. E. D.

Remark: A good wattmeter is the standard instrument for measuring power in alternating current circuits. The three-ammeter and the three-voltmeter methods are troublesome and slight errors of observation may in some cases lead to very great errors in the result.

---

* Small, that is in comparison with $\frac{r}{2\pi f}$; where r is the total resistance of a and r, Fig. 26, and f is the frequency of the alternating current.

37. The recording wattmeter is an instrument for summing up the total work or energy expended in a circuit.

The Thomson recording wattmeter is a small electric motor without iron, the field and armature coils of which constitute an electrodynamometer. The field coils B of this motor, Fig. 27, are connected in series with the circuit CC in which the work to be measured is expended.

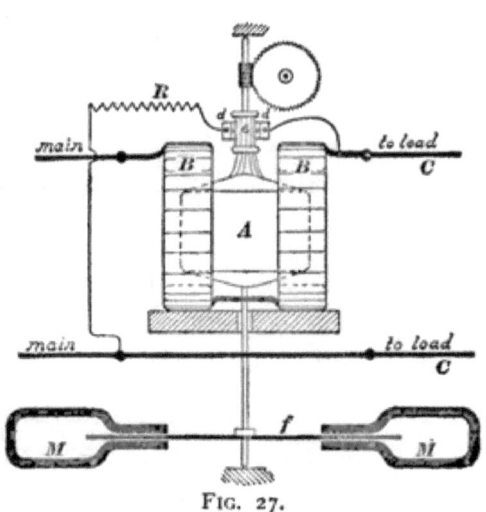

Fig. 27.

The armature A, together with an auxiliary non-inductive resistance R, is connected between the terminals of the circuit CC, as shown. Current is led into the armature by means of the brushes dd pressing on a small silver commutator e.

Discussion of the Thomson recording wattmeter.—The driving torque, acting upon the armature is proportional to the rate at which work is spent in the circuit CC (i. e., to the power expended, as explained in Art. 36). The instrument is so constructed that the speed of the armature is proportional to this driving torque or to the power spent in CC. That is, the rate of turning of the armature is proportional to the rate at which work is done in the circuit CC, so that the total number of revolutions turned by the armature is proportional to the total work expended in the circuit CC.

To make the armature speed proportional to the driving torque the armature is mounted so as to be as nearly as possible free from ordinary friction and a copper disk f, Fig. 27, is mounted on the armature spindle so as to rotate between the poles of permanent steel magnets MM. To drive such a disk requires a driving torque proportional to its speed.

CHAPTER IV.

HARMONIC ELECTROMOTIVE FORCE AND CURRENT.

38. Definition of harmonic e. m. f. and current.—A line OP, Fig. 28, rotates at a uniform rate, f revolutions per second, about a point O in the direction of the arrow gh. Consider the projection Ob of this rotating line upon the fixed line AB; this projection being considered positive when above O and negative when below O. *An harmonic e. m. f. (or current) is an e. m. f. which is at each instant proportional to the line Ob, Fig. 28.* The line Ob represents at each instant the actual value e of the harmonic e. m. f. to a definite scale; and the length of the line OP (which is the maximum length of Ob) represents the maximum value E of the harmonic e. m. f. to the same scale. The line Ob passes through a complete cycle of values during one revolution of OP and so also does the harmonic e. m. f. e. Therefore the revolutions per second f, of the line OP is the frequency of the harmonic e. m. f. e. The rotating lines E and I, Fig. 29, of which the projections on a fixed line (not shown in the figure) represent the actual instantaneous values e and i if an harmonic e. m. f. and an harmonic current are said to *represent* the harmonic e. m. f. and current respectively. Of course, the rotation of the lines E and I is a thing merely to be imagined.

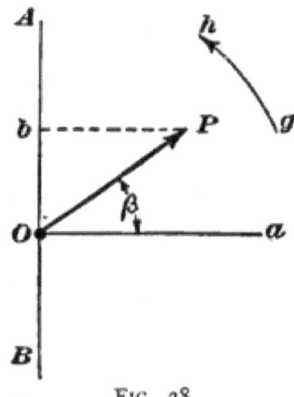

FIG. 28.

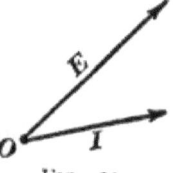

FIG. 29.

39. Algebraic expression of harmonic e. m. f. and current.—The line OP, Fig. 28, makes f revolutions per second and, therefore,

(39)

it turns through $2\pi f$ radians per second, since there are 2π radians in a revolution, that is

$$\omega = 2\pi f \qquad (28)$$

in which ω is the angular velocity of the line OP in radians per second. Let time be reckoned from the instant that OP coincides with Oa, then after t seconds OP will have turned through the angle $\beta = \omega t$ and from Fig. 28 we have

$$Ob = OP \sin \beta = OP \sin \omega t$$

But Ob represents the actual value e of the harmonic e. m. f. at the time t and OP represent its maximum value E, therefore

$$e = E \sin \omega t \qquad (29)$$

is an algebraic expression for the actual value e of an harmonic e. m. f. at time t; E being the maximum value of e, and $\dfrac{\omega}{2\pi}$ being the frequency according to equation (28).

Similarly
$$i = I \sin \omega t \qquad (30)$$

is an algebraic expression for the actual value i of an harmonic current at time t; I being the maximum value of i.

Remark 1: If time is reckoned from the instant that OP, Fig. 28, coincides with the line Ob then equations (29) and (30) become

$$e = E \cos \omega t$$
$$i = I \cos \omega t$$

Remark 2: The curve which represents an harmonic e. m. f. or an harmonic current (see Art. 21) in a curve of sines.

Remark 3: A great many alternators, especially those with distributed armature windings, generate e. m. f.'s which are very nearly harmonic. Calculations in connection with the design of alternating current apparatus are simple enough to be practicable only when the e. m. f.'s and currents are assumed to be harmonic. Hereafter, then, when speaking of alternating e. m. f.'s or currents it will be understood that they are harmonic unless it is specified to the contrary.

HARMONIC E. M. F. AND CURRENT.

40. Definitions.* *Cycle.*—A cycle is one complete set of values (positive and negative) through which an e. m. f. or current repeatedly passes. The *frequency* is the number of cycles passed through per second. The *period* is the duration of one cycle. For example, an alternator generates e. m. f. at a frequency of 60 cycles per second; the period is $\frac{1}{60}$ of a second and the angular velocity of the line OP, Fig. 28, is 60 revolutions per second or 120π radians per second ($=\omega$).

Synchronism.—Two alternating e. m. f.'s or currents are said to be in synchronism when they have the same frequency. Two alternators are said to run in synchronism when their e. m. f.'s are in synchronism.

41. Phase difference.—Consider two synchronous harmonic e. m. f.'s e_1 and e_2. Suppose that e_1 passes through its maximum value before e_2; then e_1 and e_2 are said to *differ in phase*. The line $\mathbf{E}_1$, Fig. 30, which represents e_1 must be ahead of the line $\mathbf{E}_2$ which represents e_2 as shown in the figure. The angle θ is called the *phase difference* of e_1 and e_2. The lines $\mathbf{E}_1$ and $\mathbf{E}_2$ are supposed to be rotating about O in a counter-clockwise direction as explained in Art. 38.

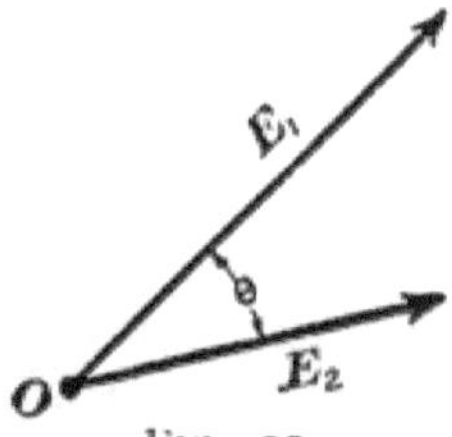

FIG. 30.

When the angle θ, Fig. 30, is zero, as shown in Fig. 31, the e. m. f.'s e_1 and e_2 are said to be *in phase*. In this case the e. m. f.'s increase together and decrease together; that is when e_1 is zero e_2 is also zero, when e_1 is at its maximum value so also is e_2, etc.

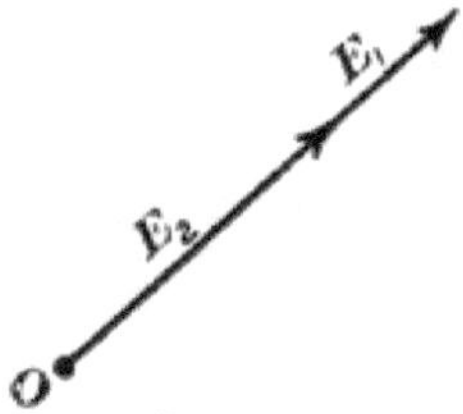

When $\theta = 90°$ as shown in Fig. 32 the two e. m. f.'s are said to be *in quadrature*. In this case one e. m. f. is zero when the other is a maximum, etc.

FIG. 31.

---

* The definitions of cycle and frequency given in Art. 20 are here repeated for the sake of clearness. All definitions given in this article apply to alternating currents and electromotive forces of any character as well as to harmonic e. m. f.'s and currents.

When $\theta = 180°$ as shown in Fig. 33 the two e. m. f.'s are said to be *in opposition*. In this case the two e. m. f.'s are at each instant opposite in sign and when one is at its positive maximum the other is at its negative maximum, etc.

42. Composition and resolution of harmonic e. m. f.'s and currents. (*a*) *Composition.*—Consider two synchronous harmonic e. m. f.'s c_1 and c_2 represented by the lines $\boldsymbol{E_1}$ and $\boldsymbol{E_2}$, Fig. 34. The sum $c_1 + c_2$ is an harmonic e. m. f. of the same frequency and it is represented by the line $\boldsymbol{E}$. This is evident when we consider that the projection on any line of the diagonal of a parallelogram is equal to the sum of the projections of the sides of the parallelogram.

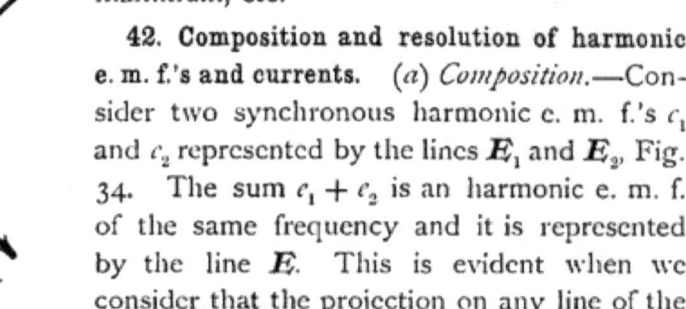

Fig. 32.

Corollary.—The sum of any number of synchronous e. m. f.'s (or currents) is another e. m. f. (or current) of the same frequency which is represented in phase and magnitude by the line which is the vector sum of the lines which represent the given e. m. f.'s (or currents). Thus the lines $\boldsymbol{E_1}$, $\boldsymbol{E_2}$ and $\boldsymbol{E_3}$, Fig. 35, represent three given synchronous harmonic e. m. f.'s and the line $\boldsymbol{E}$ (the vector sum of $\boldsymbol{E_1}$, $\boldsymbol{E_2}$ and $\boldsymbol{E_3}$) represents an harmonic e. m. f. which is the sum of the given e. m. f.'s.

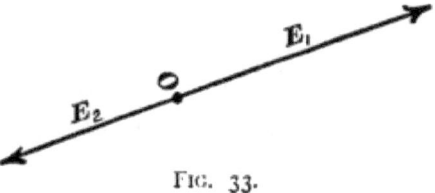

Fig. 33.

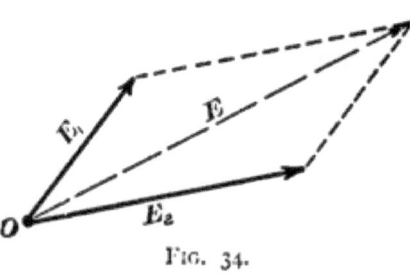

Fig. 34.

(*b*) *Resolution.*—A given harmonic e. m. f. (or current) may be broken up into a number of harmonic parts of the same frequency by reversing the process of composition. For example, the line $\boldsymbol{E}$, Fig. 35, represents a given harmonic e. m. f. which may be split up into the three e. m. f.'s represented by the lines E_1, E_2 and E_3.

43. Examples of composition and resolution.

(a) Two alternators A and B running in synchronism are connected in series between the mains as shown in Fig. 36. If

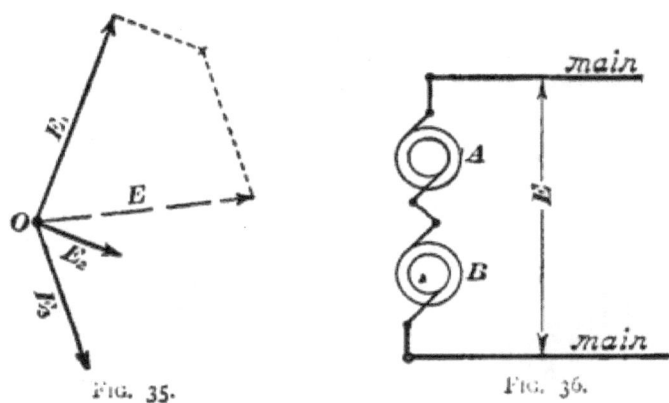

FIG. 35. FIG. 36.

the e. m. f.'s of A and B are in phase the e. m. f. between the mains will be simply the numerical sum of the e. m. f.'s of A and B. If, on the other hand, the e. m. f.'s of A and B differ in phase the state of affairs will be such as is represented in Fig. 37; in which the lines A and B represent the e. m. f.'s of the alternators A and B respectively, θ is the phase difference of A and B,

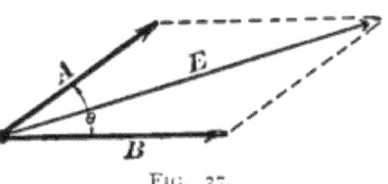

FIG. 37.

and the line E represents the e. m. f. between the mains.

(b) Two alternators A and B running in synchronism are connected in parallel between the mains shown in Fig. 38. Let the lines A and B, Fig. 39, represent the currents given by A and B re-

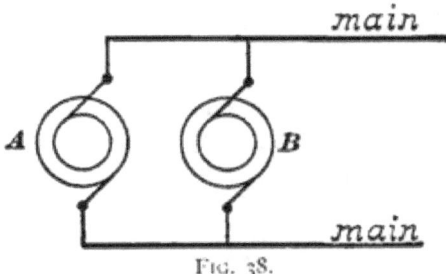

FIG. 38.

spectively, the phase difference being θ; then the current in the main line is represented by I.

(c) Two circuits A and B are connected in series between the mains of an alternator as shown in Fig. 40. The line E, Fig. 41, represents the e. m. f. between the mains, the line A represents the e. m. f. between the terminals of the circuit A and the line B represents the e. m. f. between the terminals of the circuit B. If the circuits A and B have inductance it may be that the e. m. f. A and the e. m. f. B are not in phase with each other in which case the relation between A, B and E will be as shown in Fig. 41. If one of the circuits A or B contains a con-

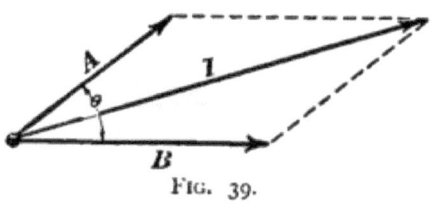

Fig. 39.

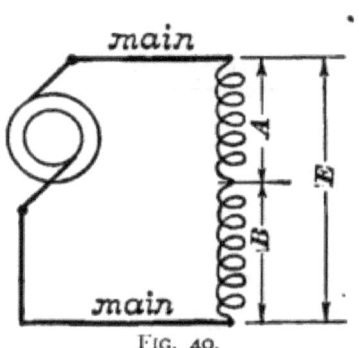

Fig. 40.

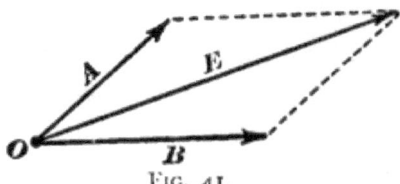

Fig. 41.

denser then the e. m. f.'s A and B, Fig. 41, may be nearly opposite in phase, and A and B may each be indefinitely greater than the e. m. f. E between the mains.

(d) Two circuits A and B, Fig. 42, are connected in parallel across the terminals of an alternator as shown. The current I from the alternator is related to the currents A and B as shown in Fig. 43. If one of the circuits A or B contains a condenser then the currents A and B may be nearly opposite in phase and the currents A and B may each be indefinitely greater than the current I from the alternator.

44. Rate of change of harmonic e. m. f.'s and currents.—Consider the harmonic current [see equation (30)]:

HARMONIC E. M. F. AND CURRENT.

$$i = I \sin \omega t \qquad (a)$$

When this current is sent through an inductive circuit an e. m. f. $L \frac{di}{dt}$ is at each instant required to make the current increase or decrease. In the study of alternating currents in inductive circuits it is, therefore, necessary to consider the rate of change $\frac{di}{dt}$ of the current.

Differentiating the above expression for i with respect to time we have

$$\frac{di}{dt} = \omega I \cos \omega t \qquad (b)$$

or

$$\frac{di}{dt} = \omega I \sin (\omega t + 90) \qquad (31)$$

FIG. 42.

This equation shows that the *rate of change,* $\frac{di}{dt}$ of an harmonic current may be represented by the projection* of the line ωI,

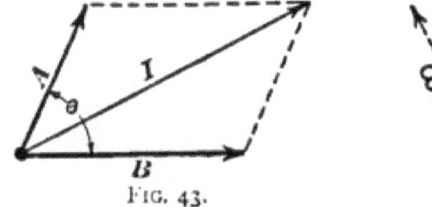

FIG. 43.

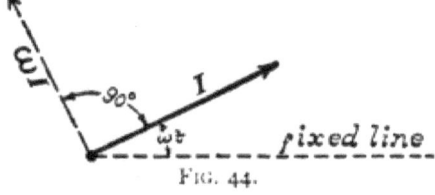

FIG. 44.

Fig. 44, which is 90° ahead of the line I which represents the current.

The relation of i and $\frac{di}{dt}$ is most clearly shown by the sine curve diagram. Thus the full line curve, Fig. 45, represents the harmonic current i. The steepness of this curve at each point represents the value of $\frac{di}{dt}$. The steepness of this curve is great-

---

*On a vertical fixed line not shown in the figure.

est at the point a where the curve crosses the axis, hence the value of $\frac{di}{dt}$ is a maximum 90° before i reaches its maximum. The ordinates of the dotted curve represent the values of $\frac{di}{dt}$.

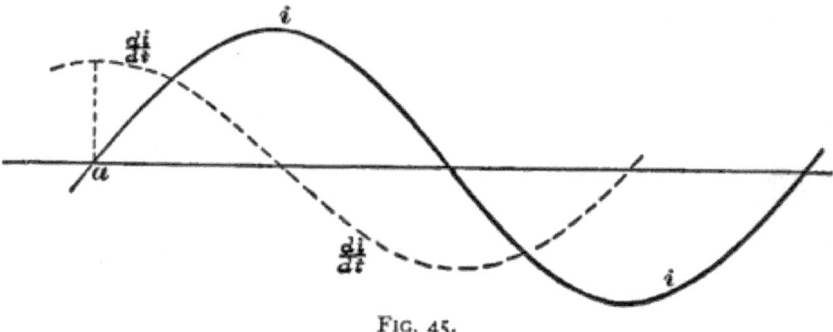

FIG. 45.

Remark: It is to be noted that the portion, $L\frac{di}{dt}$, of the total e. m. f. acting on the circuit, which is used to cause the current to increase and decrease, is represented by the line $\omega L I$, Fig. 46: the line I represents the current in the circuit.

45. Average or mean value of an harmonic e. m. f. or current.— Consider any varying quantity

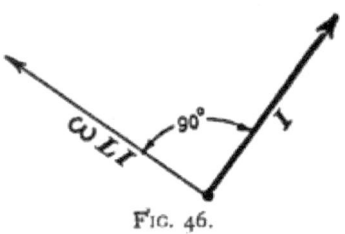

FIG. 46.

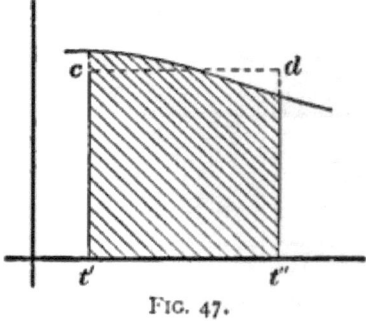

FIG. 47.

y. Its average value during an interval of time from t' to t'' is $\frac{\Sigma y \varDelta t}{t'' - t'}$, the summation being extended throughout the interval. That is,

$$\text{Av. } y = \frac{1}{t'' - t'} \int_{t'}^{t''} y\,dt \qquad (32)$$

If the successive values of y be represented by the ordinates of the curve, Fig. 47, and the corresponding values of the time t be represented by the abscissas, then $\int_{t'}^{t''} y\,dt$ is the area of the shaded portion and $\frac{1}{t''-t'}\int_{t'}^{t''} y\,dt$ is the height of a rectangle $t'cdt''$ of the same area as the curve and having the same base. The average ordinate of such a curve, as Fig. 47, during a given interval of time may be obtained quite closely by measuring the lengths of a number of equidistant ordinates. The sum of these ordinates divided by the number of ordinates gives the average ordinate of the curve.

Proposition: The average value of the sum of a number of quantities is equal to the sum of the average values of each.

Proof: Let $x, y, z, \ldots$ be the quantities. Then by definition we have

$$Av.\,(x+y+z+\ldots) = \frac{1}{t''-t'}\int_{t'}^{t''}(x+y+z\ldots)\,dt \qquad (i)$$

but

$$\frac{1}{t''-t'}\int_{t'}^{t''}(x+y+z\ldots)\,dt = \frac{1}{t''-t'}\int_{t'}^{t''} x\,dt + \frac{1}{t''-t'}\int_{t'}^{t''} y\,dt + \ldots$$

$$= Av.\,x + Av.\,y + \ldots \qquad (ii)$$

Therefore

$$Av.\,(x+y+z+\ldots) = Av.\,x + Av.\,y + Av.\,z + \ldots \qquad (33)$$

Q. E. D.

46. Proposition.—The average value of an harmonic e. m. f. or current during half a cycle $\left(\omega t = 0 \text{ to } \omega t = \pi; \text{ or } t = 0 \text{ to } t = \frac{\pi}{\omega}\right)$ is $\dfrac{2\ maximum\ value}{\pi}$.

Proof: Let $e = E \sin \omega t$ be the harmonic e. m. f. Substitute $E \sin \omega t$ for y in equation (32) and we have

$$Av.\,e = \frac{E}{t''-t'}\int_{t'}^{t''} \sin \omega t\,dt.$$

Substituting x for ωt and remembering that the limits are from $\omega t = 0$ to $\omega t = \pi$ we have

$$Av.\,e = \frac{E}{\pi}\int_0^{\pi} \sin x\,dx = \frac{E}{\pi}\Big[-\cos x\Big]_0^{\pi} = \frac{2E}{\pi} \qquad (34) \qquad \text{Q. E. D.}$$

or the average value of the harmonic e. m. f. is twice the maximum value divided by

π. Since $\frac{2}{\pi} = .636$ it may also be stated that the average value of an harmonic e. m. f., or current, is .636 times the maximum value.

Remark: The average value of an harmonic e. m. f. or current during one or more whole cycles is zero.

47. Proposition.—The square root of mean square, or effective value, of an alternating e. m. f. or current during one or more whole cycles is equal to $\dfrac{\text{maximum value}}{\sqrt{2}}$ or $.707 \times$ maximum.

Proof: Let $e = E \sin \omega t$ be a harmonic e. m. f. To find the average value of $e^2 = E^2 \sin^2 \omega t$ it is necessary to find the average value of the square of the sine of the uniformly variable angle ωt. We have the general relation

$$\sin^2 \omega t + \cos^2 \omega t = 1 \qquad (a)$$

so that by equation (33)

$$\text{Av. } \sin^2 \omega t + \text{Av. } \cos^2 \omega t = 1. \qquad (b)$$

Now, the cosine of a uniformly variable angle passes similarly through the same set of values during a cycle as the sine, hence Av. $\sin^2 \omega t$ and Av. $\cos^2 \omega t$ are equal, so that from (b) we have:

$$2 \text{ Av. } \sin^2 \omega t = 1$$

or

$$\text{Av. } \sin^2 \omega t = \tfrac{1}{2}.$$

The average value of e^2 is

$$\text{Av. } e^2 = E^2 \text{ Av.} \sin^2 \omega t$$

$$\text{Av. } e^2 = \frac{E^2}{2}$$

and

$$\sqrt{\text{Av. } e^2} = \frac{E}{\sqrt{2}} \qquad (35)$$

Q. E. D.

Note: The square root of mean square value of an harmonic e. m. f. or current is often spoken of as the *effective* value of the e. m. f. or current. When it is stated that an alternating current is so many amperes, the effective or square root of mean square value is always meant. The same is also true with regard to alternating e. m. f.'s. Hereafter the symbol E will be used to designate the effective value of an e. m. f. and I the effective value of an alternating current. *In case the currents and e. m. f.'s are harmonic* we have the relations

HARMONIC E. M. F. AND CURRENT.

$$E = \frac{E}{\sqrt{2}} \quad (36)$$

$$I = \frac{I}{\sqrt{2}} \quad (37)$$

in which E and I are the maximum values of the e. m. f. and current respectively.

Note: The ratio $\frac{\textit{effective value}}{\textit{average value}}$ of an alternating e. m. f. or current is sometimes spoken of as the *form factor* of the e. m. f. or current, because this ratio depends upon the shape or form of e. m. f. or current curve. The form factor in the case of harmonic e. m. f.'s or currents would be $\frac{.707}{.636} = 1.11$.

48. Power.—As pointed out in Article 22, Chapter II., the power developed by an alternating e. m. f. pulsates and in most practical problems it is the average power developed which is the important consideration. Let $e = E \sin \omega t$ be an harmonic e. m. f. acting on a circuit and $i = I \sin(\omega t - \theta)$ the current produced in the circuit; θ being the difference in phase of the e. m. f. and current as shown in Fig. 48.

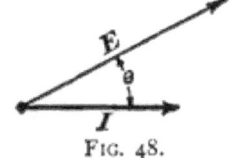

FIG. 48.

The power developed at a given instant is ei and in order to estimate the average power developed we must find an expression for the average value of ei. We have

$$ei = EI \sin \omega t \sin(\omega t - \theta)$$

or since $\sin(\omega t - \theta) = \sin \omega t \cdot \cos \theta - \cos \omega t \cdot \sin \theta$, we have

$$ei = EI \cos \theta \cdot \sin^2 \omega t - EI \sin \theta \cdot \sin \omega t \cdot \cos \omega t.$$

Hence by equation (33)

Average $ei = EI \cos \theta$ *av.* $\sin^2 \omega t - EI \sin \theta$ *av.* $\sin \omega t \cos \omega t$.

The average value of $\sin \omega t \cos \omega t$ is zero since it passes through positive and negative values alike. The average value of $\sin^2 \omega t$ is $\frac{1}{2}$. Therefore,

$$\text{Average } ei = \text{Power} = \frac{EI}{2} \cos \theta. \qquad (38)$$

It is more convenient to have this product expressed in terms of the effective values of the current and e. m. f. Hence substitute for E and I their values as given by equations (36) and (37) we have

$$\text{Power} = EI \cos \theta. \qquad (39)$$

Remark: The factor *cos θ* which depends upon the inductance and resistance of the circuit which is receiving the power [See Art. 50] is called the *power factor* of the circuit.

CHAPTER V.

FUNDAMENTAL PROBLEMS IN ALTERNATING CURRENTS.

49. Problem III.*—To determine the e. m. f. required to maintain a harmonic alternating current in a non-inductive circuit.

$$i = I \sin \omega t \qquad (a)$$

be the given harmonic current. The required e. m. f. e is used wholly to overcome the resistance R of the circuit and it is, therefore, equal to Ri so that

$$e = RI \sin \omega t \qquad (b)$$

This e. m. f. is harmonic, its maximum value is RI and it is in phase with the current i. Thus the line I, Fig. 49, represents the given harmonic alternating current and the line RI represents the e. m. f. required to maintain the given current in a non-inductive circuit.

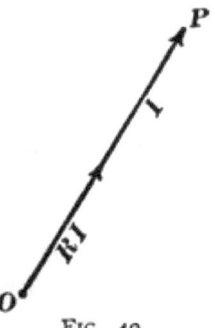

FIG. 49.

50. Problem IV.—To determine the e. m. f. required to maintain a harmonic alternating current in a circuit of resistance R and inductance L.

Let
$$i = I \sin \omega t$$
be the given harmonic current. The required e. m. f. consists of two parts, namely:

(1) The part used to overcome the resistance of the circuit.

This part is at each instant, equal to Ri; it is in phase with i and its maximum value is RI.

(2) The part used to make the current increase and decrease, or briefly to overcome the inductance. This part is, at each instant, equal to $L \dfrac{di}{dt}$ according to equation (3); it is 90° ahead of

---
* Problems I. and II. are given in Chapter I.

i in phase (see Art. 44), and its maximum value is $\omega L I$. Let the given harmonic alternating current i be represented by the line I, Fig 50. Then RI is the line which represents Ri; $\omega L I$ is the line which represents $L\dfrac{di}{dt}$; and the line E represents the total e. m. f. required to maintain the given current. From the diagram we have

$$E = I\sqrt{R^2 + \omega^2 L^2} \qquad (40)$$

in which E is the maximum value of the required e. m. f.; and further

$$\tan\theta = \dfrac{\omega L}{R} \qquad (41)$$

in which θ is the phase difference between the e. m. f. and current.

The effective value of the e. m. f. is $E = \dfrac{E}{\sqrt{2}}$, and the effective value of the current is $I = \dfrac{I}{\sqrt{2}}$ [by equations (36) and (37)] therefore substituting $\sqrt{2}\,E$ for E and $\sqrt{2}\,I$ for I in equation (40) we have

$$E = I\sqrt{R^2 + \omega^2 L^2} \qquad (42)$$

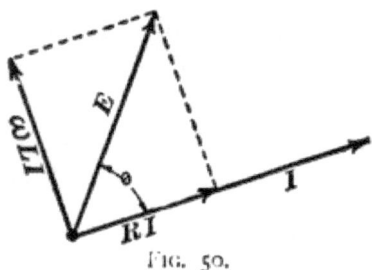

Fig. 50.

When ωL is very small compared with R the effect of inductance is negligible and this problem IV. reduces to the problem III. When ωL is very large compared with R the angle θ approaches 90° and the power $EI \cos\theta$ becomes very small, although E and I may both be considerable. In this case the current, lagging as it does 90° behind the e. m. f., is called a *wattless current*. Thus the alternating current in a coil of wire wound on a laminated iron core is approximately wattless.

Corollary: The current which is maintained in an inductive circuit by a given harmonic alternating e. m. f. is a current of which the effective value is $\dfrac{E}{\sqrt{R^2+\omega^2 L^2}}$ by equation (42) and which lags behind the e. m. f., by the angle of which the tangent is $\dfrac{\omega L}{R}$ by equation (41).

Remark: The relation between maximum values of e. m. f. and current (harmonic) is in every case the same as the relation between effective values, and henceforth effective values will, as a rule, be used in equations and diagrams. Maximum values will be indicated in the text by bold-faced letters, **E, I, N, Q**, etc.; effective values by the letters E, I, etc., and instantaneous values by e, i, n, q, etc.

51. Problem V.—To determine the current in an inductive circuit immediately after an harmonic e. m. f., $E \sin \omega t$, is connected to the circuit.

The current which can be maintained by the given e. m. f. is

$$i' = \frac{E}{\sqrt{R^2+\omega^2 L^2}} \sin(\omega t - \theta) \tag{a}$$

according to problem IV.; and the decaying current

$$i'' = C e^{-\frac{R}{L}t} \tag{10)bis}$$

can exist in the circuit independently of all outside e. m. f., C being a constant, as shown by Problem I., Chapter I. Therefore the current which can exist in an inductive circuit upon which an harmonic e. m. f. acts $i = i' + i''$ or

$$i = \frac{E}{\sqrt{R^2+\omega^2 L^2}} \sin(\omega t - \theta) + C e^{-\frac{R}{L}t} \tag{43}$$

in which e is the Napierian base, θ is the angle defined by equation (41) and C is a constant. This constant C is determined by the condition that i is equal to zero at the instant when the e. m. f. is connected to the circuit. Let t' be the given instant at which the harmonic e. m. f. begins to act upon the circuit. Substitute the pair of values $\begin{Bmatrix} t = t' \\ i = 0 \end{Bmatrix}$ in equation (43) and solve for C, the only unknown quantity; then substituting this value of C in equation (43) we have the expression for the actual current which flows in the circuit during the time that the maintained current is being established. In a very short time after the e. m. f. is connected to the circuit the second term of equation (43) disappears and the value of the current at each instant is given by the first term, which expresses the current which the given e. m. f. can maintain.

52. Problem VI.—A coil of resistance R and inductance L, and a condenser of capacity J are connected in series across alternating current mains as shown in Fig. 51. An alternating current flows back and forth through the coil and charges the condenser in one direction and the other alternately. The problem of finding the relation between the current in the coil and the e. m. f. between the mains is reduced to its simplest form as follows :

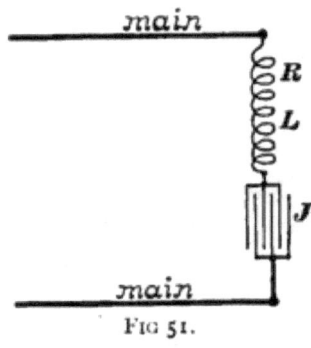

Fig 51.

To determine the e. m. f. necessary to make the charge q on the condenser vary so that

$$q = Q \sin \omega t \qquad (a)$$

in which t is elapsed time, ωt is an angle increasing at a constant rate, and Q is the maximum value of the charge in the condenser. This varying charge may be represented by the projection of the rotating line Q, Fig. 52. The current in the circuit is the rate $\dfrac{dq}{dt}$, at which the charge on the condenser changes. That is

$$i = \frac{dq}{dt} \qquad (b)$$

or from equation (a) we have

$$i = \omega Q \cos \omega t. \qquad (c)$$

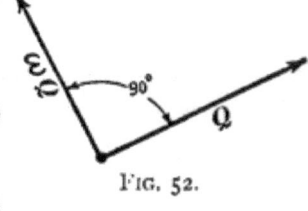

Fig. 52.

That is, the current is 90° ahead of q in phase, its maximum value is ωQ and it is represented by the line ωQ, Fig. 52. Using the symbol I for the maximum value of the current we have

$$I = \omega Q. \qquad (d)$$

The required e. m. f. is at each instant used in part to overcome the resistance R of the coil, in part to cause the current to increase and decrease in the coil, and in part to hold the charge on the condenser.

FUNDAMENTAL PROBLEMS.

1. The first part is equal to Ri at each instant. It is in phase with i, and its maximum value is RI.

2. The second part is equal to $L\dfrac{di}{dt}$ at each instant. It is 90° ahead of the current and its maximum value is ωLI.

3. The third part is equal to $\dfrac{q}{J}$* at each instant. It is in phase with q or 90° behind the current, and its maximum value is $\dfrac{Q}{J}$ or $\dfrac{I}{\omega J}$ from (d).

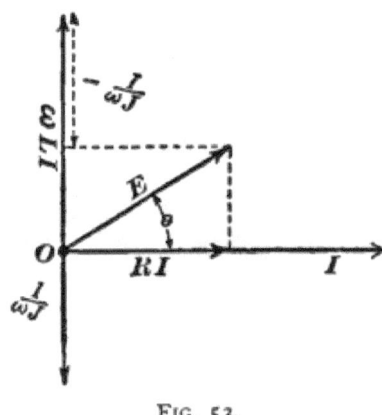

FIG. 53.

Let the line I, Fig. 53, represent the current. Then RI represents the portion of the e. m. f. used to overcome resistance, the line ωLI 90° ahead of the current represents the e. m. f. required to overcome inductance and the line $\dfrac{I}{\omega J}$ 90° behind the current represents the e. m. f. required to hold the charge on the condenser. The line E which is the vector sum of RI, ωLI and $\dfrac{I}{\omega J}$ represents the total required e. m. f. From the right triangle of which the sides are RI, $\omega LI - \dfrac{I}{\omega J}$, and E, we have

$$E = I\sqrt{R^2 + \left(\omega L - \dfrac{1}{\omega J}\right)^2}$$

or

$$E = I\sqrt{R^2 + \left(\omega L - \dfrac{1}{\omega J}\right)^2} \qquad (44)$$

and

$$\tan \theta = \dfrac{\omega L - \dfrac{1}{\omega J}}{R} \qquad (45)$$

---

* Since $q = Je$ according to equation (17), Chapter I.

Corollary.—A given harmonic e. m. f. acting upon a circuit containing a condenser of capacity J, inductance L and resistance R maintains a current of which the effective value is

$$I = \frac{E}{\sqrt{R^2 + \left(\omega L - \frac{1}{\omega J}\right)^2}} \qquad (46)$$

and which lags behind the e. m. f. by the angle θ defined by equation (45). The quantity $\omega L - \frac{1}{\omega J}$ may be either positive or negative according as ωL or $\frac{1}{\omega J}$ is the greater so that the current may be either behind or ahead of the e. m. f. in phase. In fact the limiting values of θ are $\pm 90°$.

The oscillatory current.—If $\omega L - \frac{1}{\omega J} = 0$ then the impressed e. m. f. has only to overcome the resistance of the circuit and problem VI. reduces in form to problem III. If the resistance of the circuit in this case were negligibly small then no e. m. f. at all would be required to maintain the given harmonic current. Such a self sustained harmonic current is called an *oscillatory current*. In this case from $\omega L - \frac{1}{\omega J} = 0$, we have:

$$\omega = \sqrt{\frac{1}{LJ}} \qquad (47)$$

or since $\omega = 2\pi f$ [equation (28)] we have

$$f = \frac{1}{2\pi}\sqrt{\frac{1}{LJ}}. \qquad (48)$$

This equation expresses what is called the *proper frequency* of oscillation of the inductive circuit of a condenser. In case the resistance of the circuit is not zero, which is of course the only real case, then the only current which can exist in the circuit independently of any impressed e. m. f. is a *decaying oscillatory current* the discussion of which is beyond the scope of this text.

The character of this decaying oscillatory current is shown by the curve, Fig. 54. The ordinates of this curve represent the successive values of the current produced when a charged condenser is discharged through an inductive circuit.

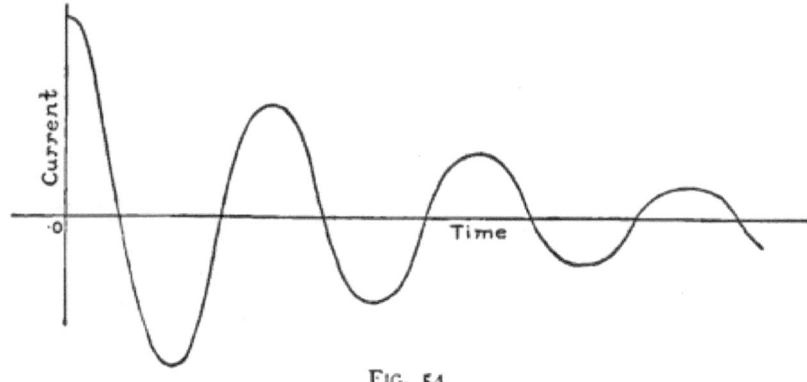

FIG. 54.

53. Electric resonance.—By inspecting equation (46) we see that an e. m. f. of given effective value E produces the greatest current in the circuit, Fig. 51 (RL and J given), when the frequency is such that $\omega L - \dfrac{1}{\omega J}$ is zero. This production of a greatest current by a given e. m. f. at a critical frequency is called

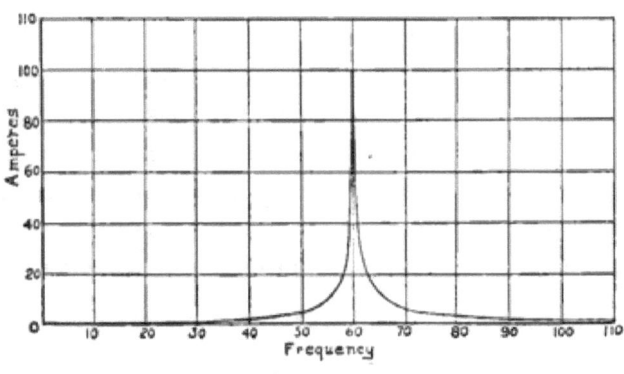

FIG. 55.

electrical resonance. Thus the ordinates of the curve, Fig. 55, represent the values of effective current at various frequencies (abscissas). The curve is based on the values $E = 200$ volts, $R = 2$ ohms, $L = .352$ henry and $J = 20$ microfarads. The maximum point of the curve is not a cusp as would appear from the figure, but the maximum is so sharply defined that it cannot be properly represented in so small a figure. When the frequency of the e. m. f. is zero, which is the case when a continuous e. m. f. acts on the circuit, the current is zero except for the very slight current which is conducted through the dielectric between the condenser plates. When the frequency of the e. m. f. is very great the current approaches zero, inasmuch as a very small current of high frequency must increase and decrease at a very rapid rate and to produce this rapid increase and decrease a very great e. m. f. is required. At low frequency the current is kept down in value by the condenser and at high frequency the current is kept down in value by the inductance.

An important point in connection with electrical resonance is that the e. m. f., $\dfrac{I}{\omega J}$, at the condenser terminals may be greatly in excess of the e. m. f. E, which is maintaining the current. Thus the line $\dfrac{I}{\omega J}$, Fig. 53, may be considerably longer than the line E.

Example: A coil of .352 henry inductance and 2 ohms resistance, and a condenser of 20 microfarads capacity, are connected in series between alternating current mains. The critical frequency of this circuit is 60 cycles per second, according to equation (48). The e. m. f. between the mains is 200 volts and its frequency is 60 cycles per second. The current in the circuit is 100 amperes according to equation (46) and the effective e. m. f. between the condenser terminals is 13,270 volts $\left(= \dfrac{I}{\omega J} \right)$.

Remark: At critical frequency $\omega L - \dfrac{1}{\omega J} = 0$ and equation (46) becomes simply $I = \dfrac{E}{R}$.

Remark: While an e. m. f. is being established between the plates of a condenser the dielectric is subjected to an increasing electrical stress and this increasing electrical stress is exactly equivalent, in its magnetic action, to an electric current flowing through the dielectric from plate to plate. Magnetically, therefore, a circuit containing a condenser is a complete circuit. Increasing (or decreasing) electrical stress is called *displacement current.*

Mechanical resonance.—The mechanical analogue of a condenser and an inductance connected in series is described in Art. 18. The body fixed to the end of a clamped flat spring (see Fig. *a*, Art. 18) will vibrate at a definite frequency when pulled to one side and released. This frequency is called the proper frequency of vibration of the body.

If a periodic force of given maximum value and given frequency acts upon the body (Fig. *a*, Art. 18) the body will be set vibrating at the same frequency as that of the force, and the violence of the motion will be greatest, for the given value of the periodic force, when the frequency of the force is equal to the proper frequency of the body. Under these circumstances the bending force (periodic) acting upon the spring may greatly exceed in value the outside force which keeps the body in motion. Thus, if a narrow strip of window glass is clamped at one end, loaded at the other end, and set vibrating by slight pushes of the finger, the strip is quickly broken if the frequency of the pushes is the same as the proper frequency of oscillation of the loaded strip. The breaking of the glass shows that the bending force acting upon it reaches values greatly in excess of the mere push of the finger.

Problems.

1. A circuit has inductance $L = 0.2$ henry and a resistance $R = 6$ ohms. Calculate the current produced by 100 volts, the frequency being 60 cycles per second. Calculate the phase dif-

ference between the e. m. f. and current. Calculate power developed.

2. A circuit has 160 ohms resistance and .2 henry inductance. Calculate the power factor of the circuit for a frequency of 60 cycles per second.

3. An e. m. f. of 20,000 volts acts on a receiving circuit of which the power factor is .85. Find the component of e. m. f. parallel to the current and the component of e. m. f. perpendicular to the current.

4. A 16-candle lamp of which the resistance is 50 ohms requires 1 ampere of current. This lamp is to be connected direct to 1,000-volt mains (frequency 133) in series with a condenser. Calculate the capacity of the condenser in order that the required current of one ampere may flow. Calculate the current when two lamps are connected in series with this condenser; when three lamps are connected in series with the condenser; and when the condenser is connected direct to the mains without any lamps in circuit.

5. An inductance of 3 henrys is connected between 1,000-volt mains (frequency 60), calculate the current in the circuit when the resistance is zero (negligibly small), when the resistance is 20 ohms, and when the resistance is 50 ohms.

6. Calculate the power taken from the mains in each case given in problems 4 and 5.

7. Two condensers of 0.5 and 0.05 microfarads capacity respectively are connected in series between 1,100-volt mains. Find the effective e. m. f. between the terminals of each condenser.

8. A quadrant electrometer of which the capacity is negligible is connected to the terminals of a 0.56-microfarad condenser. This condenser is connected between mains in series with a 0.06-microfarad condenser. The quadrant electrometer indicates 107 volts. What is the e. m. f. between the mains?

9. A quadrant electrometer has an electrostatic capacity of

.00006 microfarads. It is connected to an alternator through a wire of 50,000 ohms non-inductive resistance. Find the percentage error of the e. m. f. indications of the electrometer due to loss of e. m. f. in the resistance the frequency being 60 cycles per second.

10. The above quadrant electrometer is connected to the alternator (60 cycles per second) through a coil of 1.5 henrys inductance and of negligible resistance. Find the percentage error of the indications of the electrometer due to loss of e. m. f. in the inductance.

11. An electrodynamometer has an inductance of .0001 henry; it is to be used as a voltmeter at a frequency of 60 cycles per second. What non-inductive resistance must be connected in series with the instrument to reduce the inductance error of its readings to $\frac{1}{10}\%$.

CHAPTER VI.

THE USE OF COMPLEX QUANTITY. STEINMETZ'S METHOD.

54. Methods in alternating currents. *The graphical and the trigonometrical method.*—All fundamental problems* in alternating currents may be solved by the graphical method in which the various e. m. f.'s, currents, etc., are represented by lines in a diagram and the required results are measured off as in graphical statics. In practical problems, however, the different quantities under consideration differ so greatly in magnitude that it is difficult to scale off results with any degree of accuracy. The graphical method is, however, particularly useful for giving clear representations, and trigonometric formulas may be used in connection with graphical diagrams in every case.

The trigonometric formulæ in the more complicated problems, however, become very unwieldy and are not suitable for easily obtaining numerical results.

Steinmetz's method.—Numerical results in alternating current calculations are most easily obtained by means of a method developed mainly by Steinmetz in which *complex quantity* is used. This method is purely algebraic and is called by Steinmetz the *symbolic method.*

55. Simple quantity. Complex quantity.—A simple quantity is a quantity which depends upon a single numerical specification. Simple quantities are often called scalar quantities. A complex quantity requires two or more independent numerical specifica-

---

* The fundamental problems are those which treat of harmonic e. m. f. and harmonic current. It is a mistake to suppose that differential equations furnish a method for treating alternating currents distinct from the three methods mentioned above. In the application of differential equations the first step is to derive one or more harmonic expressions for e. m. f. and current and the subsequent development is precisely the one or the other of the above-mentioned methods.

tions to entirely fix its value. For example, if wealth depends upon the possession of horses (h) and cattle (c), then, if no agreement exists as to the relative value of horses and cattle (in fact any such agreement is essentially arbitrary), the specification of the wealth of an individual would require the specification of both horses and cattle. Thus the wealth of an individual might be $5h + 100c$. The two or more numbers which go to make up a complex quantity are called the *elements* of the quantity.

Addition and *subtraction*.—Two complex quantities are added or subtracted by adding or subtracting the similar numerical elements of the quantities. Thus $5h + 100c$ added to $6h + 15c$ gives $11h + 115c$ and $2h + 25c$ subtracted from $5h + 100c$ gives $3h + 75c$.

Multiplication and *division*.—Consider two complex quantities $5h + 2c$ and $3h + 4c$ in which h and c are independent incommensurate units, say horses and cattle. Multiplying the first of these expressions by the latter, *using the ordinary rules of algebra*, we have :

$$(5h + 2c)(3h + 4c) = 15h^2 + 6hc + 20ch + 8c^2.$$

Now in general the squares and products of units c^2, h^2, ch and hc have no meaning ; so that the significance of the product of two complex quantities depends upon *arbitrarily chosen meanings for these squares and products of units*.

56. Vectors.—A vector is a quantity which has both magnitude and direction. A vector may be specified by giving its components in the direction of arbitrarily chosen axes of reference. In specifying a vector by its components, it is necessary to have it distinctly stated which is its x component and which is its y component.* This may be done either by verbal statement or by marking one of the components by a distinguishing index. Further it is allowable to connect the two components with the sign of addition. Thus $a + jb$ is a specification of the vector of which the x component is a and the y component is b ; the

---

* We are at present concerned only with vectors in one plane.

index letter j being used to mark the y component. This expression of a vector is a complex quantity the independent units of which are vertical and horizontal distances or components.

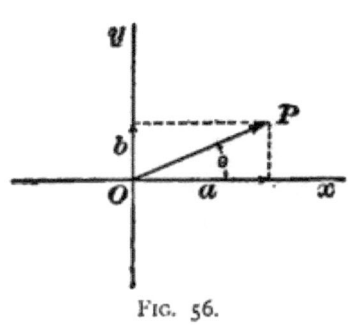

FIG. 56.

For example, the vector OP, Fig. 56, is specified by the horizontal component a and the vertical component b. The vector may therefore be specified by the expression $a + jb$ the index j being used to show that b is the vertical component.

Numerical value and direction of a vector.—The numerical value of a vector is the square root of the sum of the squares of its components. Thus the numerical value of the vector $a + jb$ is $\sqrt{a^2 + b^2}$.

The angle between the vector and the x axis is the angle, θ, of which the tangent is $\dfrac{b}{a}$, that this $\tan \theta = \dfrac{b}{a}$. This matter of value and direction of a vector is an important consideration in alternating current problems.

Addition and subtraction of vectors.—The sum of a number of vectors is a vector of which the x component is the sum of the x components of the several vectors, and of which the y component is the sum of the y components of the several vectors. Thus the sum of the vectors $a + jb$, $a' + jb'$, $a'' + jb''$ is

$$(a + a' + a'') + j(b + b' + b'').$$

The difference of the two vectors $a + jb$, $a' + jb'$ is

$$(a - a') + j(b - b').$$

Multiplication of vectors.—Consider the two vectors $a + jb$ and $a' + jb'$. Multiply these two expressions, using the formal rules of algebra, and we have for the product

$$aa' + jab' + ja'b + j^2bb'.$$

Each term in this product must be interpreted arbitrarily or according to convention in order that this product may have a definite meaning.

In the first place, we may take aa', which is not affected by the index j, to be a horizontal quantity and we may take ab' and $a'b$ to be vertical quantities or vertical components of a vector. As to the term j^2bb' we may note that the index letter j, used once, indicates that a quantity is vertical, while without the index j the quantity would be understood to be horizontal and to the right (see Fig. 56). That is, the letter j may be thought of as turning a quantity through 90° in the positive direction, that is, counter clockwise. It is, therefore, convenient to think of the letter j when used twice ($jjbb'$ or j^2bb') as turning the quantity upon which it operates through 180° or as reversing its direction and, therefore, its algebraic sign. That is, j^2bb' is to be taken as equal to $-bb'$ or

$$j^2 = -1$$

so that the product of the two vectors $a + ja'$ and $b + jb'$ is to be interpreted as the vector

$$(ab - a'b') + j(ab' + a'b).$$

Quotient of two vectors.—Consider the quotient

$$\frac{a + jb}{a' + jb'}$$

multiply both numerator and denominator by $a' - jb'$, remembering that $j^2 = -1$, and we have

$$\frac{a + jb}{a' + jb'} = \frac{aa' + bb'}{a'^2 + b'^2} + j\left(\frac{a'b - ab'}{a'^2 + b'^2}\right)$$

which leads to the conception of the quotient of two vectors as a vector of which the x component is

$$\frac{aa' + bb'}{a'^2 + b'^2}$$

and the y component is

$$\frac{a'b - ab'}{a'^2 + b'^2}.$$

We shall now apply the symbolic method to the discussion of Problems IV. and VI. in order to illustrate its use and to point out a few definitions. In the next chapter this method will be applied to more complicated problems.

57. Application of the symbolic method. *Problem IV. again.*—Consider an alternating current I (effective value), in a circuit of which the resistance is R and the inductance is L. Let the line I, Fig. 57, which represents the current, be chosen as the x-axis of reference. Let E be the e. m. f. which is maintaining the current. The x-component of E is RI and the y-component of E is ωLI. Therefore:

$$E = (R + j\omega L)I. \qquad (49)$$

Problem VI. again.—In this case, as shown in Fig. 58, the x-component of E is RI and the y-component of E is $\omega LI - \dfrac{I}{\omega f}$.

Therefore:
$$E = \left[R + j\left(\omega L - \frac{1}{\omega f}\right)\right]I. \qquad (50)$$

Impedance, resistance, reactance.—The complex quantity by which I is multiplied to give E is called the *impedance* of a circuit. Thus the impedance of the circuit discussed in problem

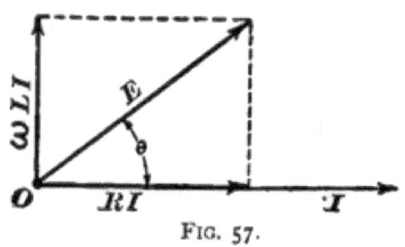
Fig. 57.

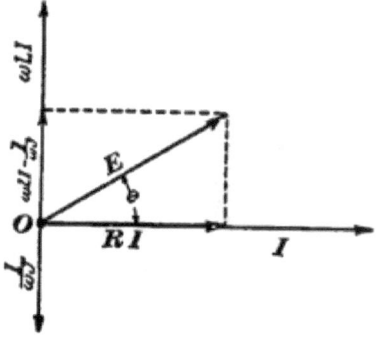
Fig. 58.

IV. is $R + j\omega L$, and the impedance of the circuit discussed in problem VI. is $R + j\left(\omega L - \dfrac{1}{\omega f}\right)$.

The quantity by which the current is multiplied to give the component of E parallel to the current is called the resistance of the circuit. Thus the resistance of the circuits discussed in Problems IV. and VI. is R.

The quantity by which the current is multiplied to give the component of E perpendicular to the current is called the reactance of the circuit. Thus the reactance of the circuit discussed in problem IV. is ωL and the reactance of the circuit discussed in problem VI. is $\omega L - \dfrac{1}{\omega f}$. The reactance due to a condenser is negative. Reactance is frequently represented by the letter x. It is to be kept in mind that reactance depends upon the frequency $f \left(= \dfrac{\omega}{2\pi} \right)$ as well as upon inductance and capacity.

Admittance. Conductance. Susceptance. Consider a circuit of resistance r and reactance x in which an alternating current I is maintained by an e. m. f. E. Then

$$E = (r + jx) I$$

or
$$I = \dfrac{1}{r + jx} \cdot E$$

or
$$I = \dfrac{r - jx}{r^2 + x^2} \cdot E$$

or
$$I = \left(\dfrac{r}{r^2 + x^2} - j \dfrac{x}{r^2 + x^2} \right) E. \quad (51)$$

The quantity, $\dfrac{r}{r^2 + x^2}$, by which E is multiplied to give the component of I parallel to E is called the *conductance* of the circuit.

The quantity, $\dfrac{x}{r^2 + x^2}$, by which E is multiplied to give the component of I perpendicular to E is called the *susceptance* of the circuit.

Equation (51) is sometimes written

$$I = (g - jb) E \quad (52)$$

in which
$$g = \dfrac{r}{r^2 + x^2} \quad (53)$$

and
$$b = \dfrac{x}{r^2 + x^2} \quad (54)$$

The complex quantity $g - jb$ is called the *admittance* of the circuit.

Remark: Resistance and reactance are both expressed in ohms, and the numerical value of impedance, which is the square root of the sum of the squares of resistance and reactance, is expressed in ohms. Thus a circuit of which the inductance is .02 henry, has a reactance of 7.54 ohms at a frequency of 60 cycles per second; a condenser of which the capacity is 2 microfarads has a negative reactance of 1325 ohms at a frequency of 60 cycles per second.

Problems.

1. Separate the components of the complex expression

$$I = \frac{a + jb}{a' + jb'}$$

2. A circuit carrying current at a given frequency has a resistance of 5 ohms and a reactance of 10 ohms. Ten amperes of current are flowing. Calculate the components of the e. m. f. and calculate the full effective value of the e. m. f.

3. At what frequency is the reactance due to .352 henry inductance equal (but, of course, opposite in sign) to the reactance of a 20-microfarad condenser?

4. Separate the components of the complex expression

$$E = r_1 I_1 + j\omega L_1 I_1 + \frac{\omega^2 M^2 I_1}{r_2 + j\omega L_2}$$

CHAPTER VII.

FURTHER FUNDAMENTAL PROBLEMS.

58. Problem VII. Coils in series. *Electromotive force consumed in transmission lines.*—An e. m. f. E acts upon two coils in series as shown in Fig. 59. For example, one coil may rep-

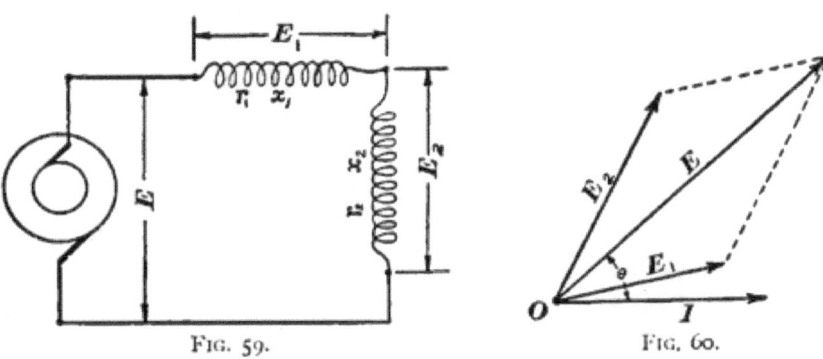

FIG. 59.   FIG. 60.

resent a transmission line and the other a receiving circuit; the total resistance and reactance of the two transmission lines being looked upon as located in one, only, of the coils for the sake of simplicity. It is required to find E_1 and E_2 each in terms of E, r_1, r_2, x_1 and x_2. Let I be the current in the circuit. The general relation between E, E_1, E_2 and I is shown in Fig. 60. The symbolic method, however, affords the easiest and simplest solution of the problem as follows:

$$E = E_1 + E_2 \tag{a}$$

This is a vector equation and expresses the fact that E is the resultant of E_1 and E_2 as shown in Fig. 60

$$E_1 = (r_1 + jx_1) I \tag{b}$$

$$E_2 = (r_2 + jx_2) I. \tag{c}$$

These equations (b) and (c) come from problem IV., Chapter VI. From equations (a), (b) and (c) we have

$$I = \frac{E}{r_1 + r_2 + j(x_1 + x_2)}. \qquad (d)$$

Substituting this value of I in (b) and (c) we have

$$E_1 = E\frac{r_1 + jx_1}{r_1 + r_2 + j(x_1 + x_2)} \qquad (e)$$

and

$$E_2 = E\frac{r_2 + jx_2}{r_1 + r_2 + j(x_1 + x_2)} \qquad (f)$$

These equations (e) and (f) express the e. m. f.'s E_1 and E_2 in terms of the known quantities E, r_1, r_2, x_1 and x_2. For purposes of numerical calculation (e) [and likewise (f)] must be separated into components, that is, into real and imaginary parts, and the numerical value of E_1 (and likewise of E_2) is then found by taking the square root of the sum of the squares of these components. Thus, multiplying numerator and denominator of (e) by $r_1 + r_2 - j(x_1 + x_2)$ we remove j from the denominator and may then separate the components, namely

$$E_1 = E\frac{r_1(r_1 + r_2) + x_1(x_1 + x_2)}{(r_1 + r_2)^2 + (x_1 + x_2)^2} + jE\frac{x_1(r_1 + r_2) - r_1(x_1 + x_2)}{(r_1 + r_2)^2 + (x_1 + x_2)^2}. \qquad (g)$$

The first term of this expression is the component of E_1 parallel to E and the second term, dropping j, is the component of E_1 perpendicular to E and the numerical value of E_1 is the square root of the sum of the squares of these components.* The final result is so complicated that it is of no use whatever in giving a conception of the phenomenon under consideration. It is useful only when it is desired to carry out numerical calculations. It is, indeed, generally the case that the use of the symbolic method in the solution of alternating current problems is simple and instructive in its initial steps only, while the final solution

---

* E_1 (numerical value) $= \dfrac{\sqrt{2r_1^2(r_1 + r_2)^2 + 2x_1^2(x_1 + x_2)^2}}{(r_1 + r_2)^2 + (x_1 + x_2)^2} E$.

itself is unintelligible. The final results will, therefore, be written out in full only when the student is expected to use them in numerical calculations.

The following simple cases of the problem under consideration are particularly interesting.

1. When $x_1 = 0$ and $x_2 = 0$; then E, E_1, E_2, and I are parallel and $E = E_1 + E_2$ (numerically).

2. When $\frac{x_1}{r_1} = \frac{x_2}{r_2}$; then E, E_1, and E_2 are parallel to each other and all are ahead of I in phase, by the angle of which the tangent is $\frac{x_1}{r_1}$. In this case, also, $E = E_1 + E_2$ (numerically).

3. When $\frac{x_1}{r_1}$ is very small and $\frac{x_2}{r_2}$ very large; then E_1 is parallel to I and E_2 is at right angles to I, as shown in Fig. 61. This figure is, of course, a particular case of Fig. 60. In the present case the numerical relation between E, E_1 and E_2 is $E = \sqrt{E_1^2 + E_2^2}$; and when E_1 is small E_2 is sensibly equal to E numerically.

Example: A transmission line of large resistance r_1 and small reactance x_1 supplies current to a receiving circuit of large reactance x_2 and small resistance r_2, so that $\frac{x_1}{r_1}$ is very small and $\frac{x_2}{r_2}$ is very large. The e. m. f. E_1, Fig. 61, consumed in the line is due almost wholly to resistance and if E_1 is not very large then E_2 is very nearly equal to E. That is the *resistance drop* in a transmission produces but

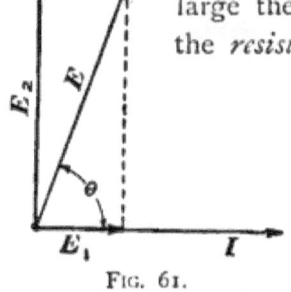
FIG. 61.

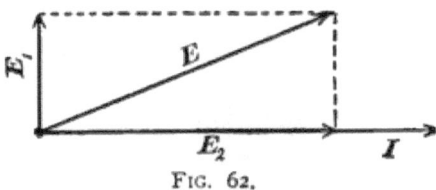
FIG. 62.

very little diminution of e. m. f. at the terminals of a receiving circuit of large reactance.

4. When $\frac{x_1}{r_1}$ is very large and $\frac{x_2}{r_2}$ very small the state of affairs is shown in Fig. 62, and $E = \sqrt{E_1^2 + E_2^2}$.

Example: A transmission line of large reactance x_1 and small resistance, r_1, supplies current to a receiving circuit of large resistance r_2 and small reactance x_2. The e. m. f., E_1, Fig. 62, consumed in the line is due almost wholly to reactance and if E_1 is not very large then E_2 is very nearly equal to E. That is the *reactance drop* in a transmission line affects the e. m. f. at the terminals of a large resistance receiving circuit but little.

5. When $\frac{x_1}{r_1}$ is large and positive and $\frac{x_2}{r_2}$ is large, but negative. Then E_1 is nearly 90° ahead of I and E_2 is nearly 90° behind I, as shown in Fig. 63. The figure shows the limiting case for which $\frac{x_1}{r_1} = +\infty$ and $\frac{x_2}{r_2} = -\infty$. In this case the total e. m. f. E is numerically equal to the difference between E_1 and E_2.

Examples: (a) A transmission line of small resistance r_1 and large reactance x_1 supplies current to a condenser. The state of affairs is shown in Fig. 63 and the e. m. f. E_2 at the terminals of the condenser exceeds the generator e. m. f. E by the amount E_1. Therefore reactance drop in a transmission line increases the e. m. f. at the terminals of a receiving circuit of negative reactance.

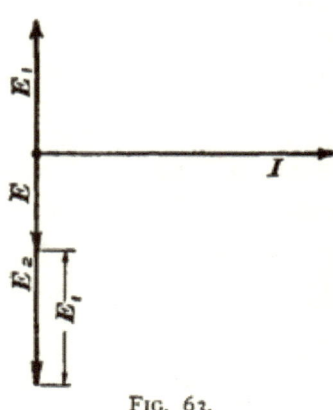

Fig. 63.

(b) A transmission line of small resistance r_1 and large reactance x_1 supplies current to a synchronous motor running at light load; e. m. f. of motor being less than e. m. f. of generator. In this case the e. m. f. at the terminals of the receiving circuit is nearly 90° behind the current in phase, as in case of the condenser, and the e. m. f. at the receiving circuit terminals is increased by the reactance drop in the line.

(c) An inductance is connected in series with a condenser between alternating current mains. If the resistance of the circuit is comparatively small the e. m. f. at the condenser terminals is nearly equal to the sum of the e. m. f. between mains plus the e. m. f. between the terminals of the inductance. (See Arts. 52 and 53.)

59. Problem VIII. Coils in parallel.—A given alternating current I divides between two circuits in parallel, as shown in Fig. 64. It is required to find I_1 and I_2 each in terms of I, r_1, r_2, x_1, and x_2. Let E be the e. m. f. between the branch points. The

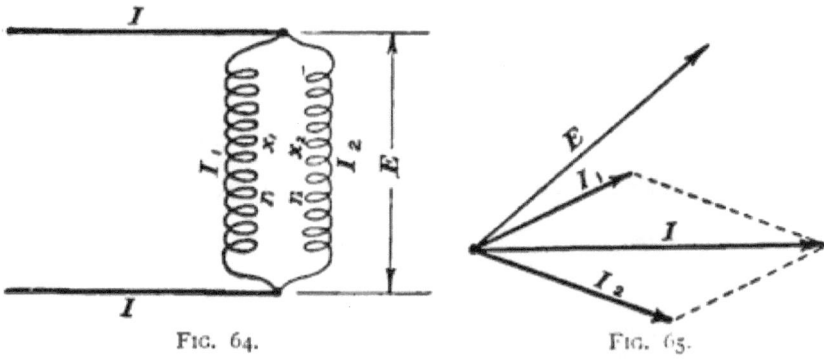

FIG. 64. FIG. 65.

general relation between I, I_1, I_2, and E is shown in Fig. 65. The symbolic method, however, affords the easiest and simplest solution of the problem as follows:

$$I = I_1 + I_2 \tag{a}$$

$$I_1 = \frac{E}{r_1 + jx_1} \tag{b}$$

$$I_2 = \frac{E}{r_2 + jx_2} \tag{c}$$

whence

$$I = E\left(\frac{1}{r_1 + jx_1} + \frac{1}{r_2 + jx_2}\right)$$

or

$$E = I\,\frac{r_1 r_2 - x_1 x_2 + j(r_1 x_2 + r_2 x_1)}{r_1 + r_2 + j(x_1 + x_2)}. \tag{d}$$

74   THE ELEMENTS OF ALTERNATING CURRENTS.

Substituting this value of E in (b) and (c) we have

$$I_1 = I \frac{r_1 r_2 - x_1 x_2 + j(r_1 x_2 + r_2 x_1)}{r_1(r_1 + r_2) - x_1(x_1 + x_2) + j[r_1(x_1 + x_2) + x_1(r_1 + r_2)]} \quad (e)$$

and a similar expression for I_2. In this expression for I_1 the index letter j may be removed from the denominator by multiplying numerator and denominator by

$$r_1(r_1 + r_2) - x_1(x_1 + x_2) - j[r_1(x_1 + x_2) + x_1(r_1 + r_2)]$$

when the components of I_1 parallel to I and perpendicular to I may be separated exactly as in problem VII.

The following simple cases of the problem under consideration are interesting.

1. When $x_1 = 0$ and $x_2 = 0$; then I, I_1, I_2 and E are parallel and $I = I_1 + I_2$ (numerically).

2. When $\frac{x_1}{r_1} = \frac{x_2}{r_2}$ then I, I_1, and I_2 are parallel to each other and all behind E in phase by the angle of which the tangent is $\frac{x_1}{r_1}$. In this case, also, $I = I_1 + I_2$ (numerically).

3. When $\frac{x_1}{r_1}$ is very small and $\frac{x_2}{r_2}$ is very large or *vice versa*. In this case I_1 is parallel to E and I_2 is 90° behind E or *vice versa* as shown in Figs. 66 and 67. These figures are particular cases

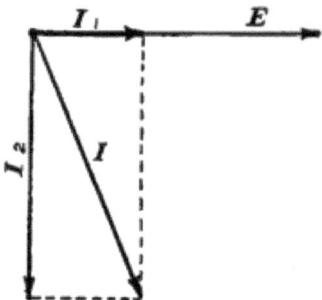

FIG. 66.

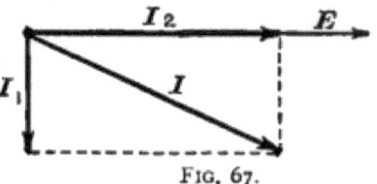

FIG. 67.

of Fig. 65. In the present case the numerical relation between I, I_1, and I_2 is $I = \sqrt{I_1^2 + I_2^2}$ and when either I_1 or I_2 is small the other is sensibly equal (numerically) to I.

4. When $\frac{x_1}{r_1}$ is large and positive and $\frac{x_2}{r_2}$ large but negative then I_1 is 90° behind E, and I_2 is 90 degrees ahead of E, as shown in Fig. 68. In this case $I = I_2 - I_1$ (numerically), that is, the sum of the currents in the branches exceeds the total current I.

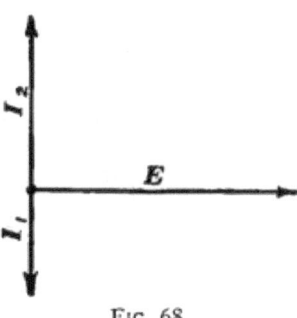

Fig. 68.

Examples of case 4: A condenser and an inductance are connected in parallel in an alternating current circuit. The current I in the circuit divides in the two branches formed by the inductance and the condenser. The currents I_1 and I_2 in these two branches are nearly opposite to each other in phase and the current I is numerically equal to the difference of I_1 and I_2.

If the frequency of the alternating current I is such that ωL and $\frac{1}{\omega J}$ are equal then I_1 and I_2 will be nearly equal; and I_1 and I_2 will each be very much greater than I.

60. Problem IX. *Compensation for lagging currents.*—The current I_2, Fig. 69, in a receiving circuit lags behind the e. m. f.,

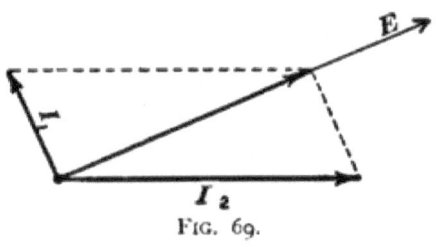

Fig. 69.

E which acts upon the terminals of the receiving circuit. If an auxiliary circuit having negligible resistance and negative reactance is connected between the terminals of the receiving circuit (*i. e.,* in parallel with the receiving circuit), then this auxiliary circuit will take the current I_1 which is 90° ahead of E, and the total current given by the alternator will be reduced to I in phase with E.

Discussion: From problem IV. we have

$$I_2 = \frac{E}{r_2 + jx_2} \tag{a}$$

or

$$I = E\left(\frac{r}{r_2^2 + x_2^2} - j\frac{x_2}{r_2^2 + x_2^2}\right) \tag{b}$$

so that the component of I perpendicular to E is $-E\dfrac{x_2}{r_2^2 + x_2^2}$. Further, if the auxiliary circuit has no perceptible resistance and x_1 is its reactance then the numerical value of I_1 is $\dfrac{E}{x_1}$ and in order that I may be parallel to E we have

$$-\frac{Ex_2}{r_2^2 + x_2^2} = \frac{E}{x_1}$$

or

$$x_1 = -\frac{r_2^2 + x_2^2}{x_2}$$

which expresses the value of the reactance (negative) which the auxiliary circuit must have to compensate for the lagging current in the receiving circuit.

Example: An alternator giving 2000 volts at a frequency of 133 cycles per second ($\omega = 2\pi \times 133$) furnishes current to a circuit of which the resistance r_2 is 100 ohms and the inductance is .03 henry. The reactance ωL of the circuit is therefore $2\pi \times 133 \times .03$ or 25 ohms. The value of $x_1 \left(=\dfrac{1}{\omega J}\right)$ required to compensate for lagging current in this circuit is 425 ohms so that the capacity J of the condenser required for compensation is .0000028 farad or 2.8 microfarads. The synchronous motor running at light load has negative reactance and may be used in place of a condenser.

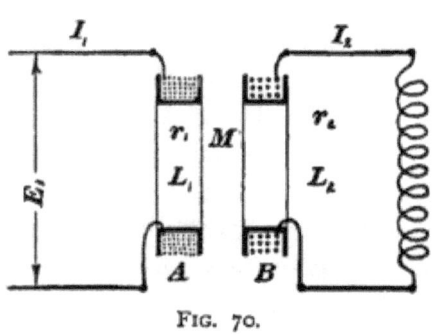

FIG. 70.

61. Problem X. The transformer without iron.—Given two coils of wire, Fig. 70, a primary coil A and a secondary coil B placed near together but not electrically connected. The secondary coil B is connected to a closed circuit of which the total resistance is r_2 and the total inductance is L_2. The resistance of the primary coil is r_1 its inductance is L_1 and the mutual inductance of the two coils is M. It is required to find the e. m. f. which must act upon the terminals of the primary coil to maintain in this coil a given harmonic current.

Determination of the secondary current I_2.—The given primary current induces in the secondary coil an e. m. f. which is at each instant equal to $M\dfrac{di_1}{dt}$.*

This e. m. f. is 90° ahead of I_1, its effective value is $\omega M I_1$, and its symbolic expression is $j\omega M I_1$, so that according to problem IV. this e. m. f. produces in the secondary coil a current

$$I_2 = \frac{j\omega M I_1}{r_2 + j\omega L_2}. \tag{a}$$

Reaction of secondary current upon the primary coil.—The secondary current I_2 induces in the primary coil an e. m. f. which is at each instant equal to $M\dfrac{di_2}{dt}$. This e. m. f. is 90° ahead of I_2 in phase, its effective value is $\omega M I_2$, and its complex expression is $j\omega M I_2$. This e. m. f. induced in the primary by the secondary current must be overcome by the e. m. f. which acts upon the primary. The portion of the acting e. m. f. which thus balances the reaction of the secondary current is equal to this reaction and opposite to it in sign and is, therefore, equal to $-j\omega M I_2$.

Determination of total e. m. f. acting on primary.—This total e. m. f. consists of three parts as follows:

(1) The part described above which balances the reaction of the secondary current. This part is equal to $-j\omega M I_2$ or using

---

* The expression for this induced e. m. f. is usually written $-M\dfrac{di}{dt}$; this negative sign is, however, a conventional matter.

the value of I_2 from equation (a) we have for this part of the total e. m. f.

$$+\frac{\omega^2 M^2 I_1}{r_2 + j\omega L_2}.$$

(2) The part used to overcome the resistance of the primary coil. This is at each instant equal to $r_1 i_1$, its effective value is $r_1 I_1$, and its complex expression is $r_1 I_1$ since it is in phase with I_1.

(3) The part used to overcome the inductance of the primary coil. This is at each instant equal to $L_1 \frac{di_1}{dt}$, it is 90° ahead of I_1, its effective value is $\omega L_1 I_1$, and its complex expression is $j\omega L_1 I_1$. Therefore the total e. m. f. required to maintain the given primary current is

$$E_1 = r_1 I_1 + j\omega L_1 I_1 + \frac{\omega^2 M^2 I_1}{r_2 + j\omega L_2} \qquad (b)$$

or separating components

$$E_1 = \left[\left(r_1 + \frac{r_2 \omega^2 M^2}{r_2^2 + \omega^2 L_2^2}\right) + j\left(\omega L_1 - \frac{\omega^3 L_2 M^2}{r_2^2 + \omega^2 L_2^2}\right)\right] I_1 \qquad (c)$$

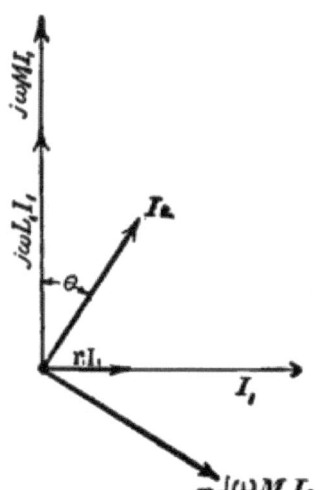

Fig. 71.

Fig. 71 shows the primary current I_1, the e. m. f. $j\omega M I_1$ induced in the secondary coil, the secondary current I_2, the portion of the primary e. m. f. $-j\omega M I_2$ used to balance the reaction of the secondary current, the portion of the primary e. m. f. $r_1 I_1$ used to overcome primary resistance, and the portion of the primary e. m. f. $j\omega L_1 I_1$ used to overcome primary inductance. The total primary e. m. f. is the vector sum of $-j\omega M I_2$, $r_1 I_1$ and $j\omega L_1 I_1$.

Equation (c) shows that the effect of the secondary coil is to make the primary coil behave as if its resistance were increased by the amount $\frac{r_2 \omega^2 M^2}{r_2^2 + \omega^2 L_2^2}$, and its inductance decreased by the amount $\dfrac{L_2\omega^2 M^2}{r_2^2 + \omega^2 L_2^2}$.

Problems.

1. A transmission line having an inductance of .02 henry and a resistance of 25 ohms supplies current at a frequency of 60 cycles per second to condenser of which the capacity is 0.24 microfarad. The e. m. f. of the alternator generator is 20,000 volts. Calculate the effective and maximum e. m. f. at the condenser terminals.

2. A transformer (without iron) consists of two long cylindrical coils each having 10 turns of wire per centimeter of length (one layer). The coils are each 50 cm. long and their radii are 2 cm. and 3 cm. respectively, the smaller coil being inside the larger. Calculate the value and phase of the e. m. f. required to maintain a current of 10 amperes at 60 cycles per second in the outer coil; calculate the current in the inner coil, and calculate the apparent reactance and resistance of the outer coil. The outer coil has 2 ohms resistance and the inner coil has 1½ ohms resistance and its terminals are short circuited.*

3. A 16-candle 50-volt lamp has about 50 ohms resistance. Such a lamp is connected in series with an inductance of 2 henrys, and another is connected in series with a condenser of which the capacity is 3 microfarads. These two circuits are connected in parallel, and through a third lamp to 500-volt 125-cycle mains. Calculate the current in each lamp.

---

* The mutual inductance, in henrys, of two coaxial solenoids is, approximately,

$$M = 4\pi^2 r''^2 z' z'' l \div 10^9,$$

in which z' and z'' are the turns of wire per unit length on the respective coils, r'' is the radius of the inside coil, and l is the length of the coils.

CHAPTER VIII.

SINGLE-PHASE AND POLYPHASE ALTERNATORS.*

62. Advantages of alternating currents.—When electric power is transmitted over considerable distances the transmission wires must be very large to prevent excessive loss, unless the power is transmitted by a small current pushed by a high e. m. f. or pressure. Considerations of safety to life and property, however, make it inadvisable to furnish power to the user in the form of small current at high pressure, so that it is now usual to transform the transmitted power and furnish it to the user in the form of large current at low pressure. The advantage of alternating currents over direct currents lies mainly in the simplicity and comparative cheapness of the alternating current apparatus† required to transform from a high pressure and small current to a low pressure and large current or *vice versa*. Alternating current generators also are better adapted for the generation of high potential than direct current generators on account of the absence of the commutator and the comparative simplicity of the armature winding which admits of high insulation.

63. The single-phase alternator and its limitations.—The simple alternator described in Chapter II. is called a *single-phase* alternator. It has one pair of collector rings to which the terminals of the armature winding are connected. The current given by a single-phase alternator is entirely satisfactory for electric lighting‡ and in general for all purposes in which the heating

---

* The single-phase alternator has been described in Chapter II. The present chapter deals with the essential features of alternators for producing polyphase currents and Chapter IX. treats in detail of the theory and design of single-phase and polyphase alternators. The meaning of the term polyphase will appear in the course of the following discussion.

† See Chapter X. The alternating-current transformer.

‡ Arc lamps do not operate quite as well with alternating current as with direct current.

effect only of the current is important. For motive purposes the simple alternating current is not satisfactory, as it is difficult to make a single-phase alternating-current motor which will start satisfactorily under load. For electrochemical processes the alternating current cannot be used.

The satisfactory use of alternating currents for motive purposes depends mainly upon the use of the *induction motor* described in Chapter XIII. It is the requirements of this motor which has led to the development of polyphase systems.

64. The two-phase alternator.—Consider two similar and independent single-phase armatures A and B, Fig. 72, mounted rigidly on the same shaft, one beside the other, and revolved inside the same crown of field magnet poles. In the figure, armature B is shown inside of A for the sake of clearness. These armatures are so mounted on the shaft that the slots of A are midway under the poles when the slots of B are midway between the poles as shown. Under these conditions the e. m. f.'s of A and B are so related in their pulsations that the e. m. f. of A is at its maximum when the e. m. f. of B is zero, that is the e. m. f.'s are 90° apart in phase, or in quadrature with each other. Two alternators connected (mechanically) in the manner indicated constitute a *two-phase alternator*. The two distinct and independent e. m. f.'s generated by such a machine are used to supply two distinct and independent currents to two distinct and independent circuits. In practice the two-phase alternator is made by placing the armature windings of A and B upon one

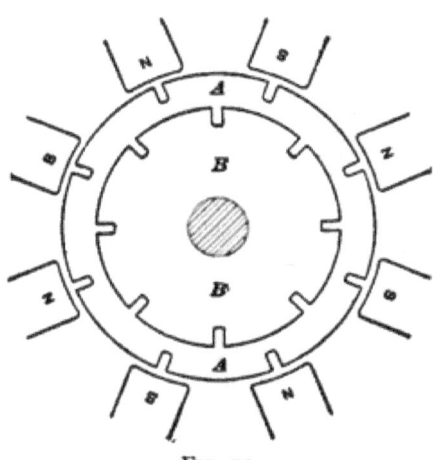

FIG. 72.

and the same armature body. For this purpose the armature body has twice as many slots as A or B, Fig. 72. Fig. 73 shows

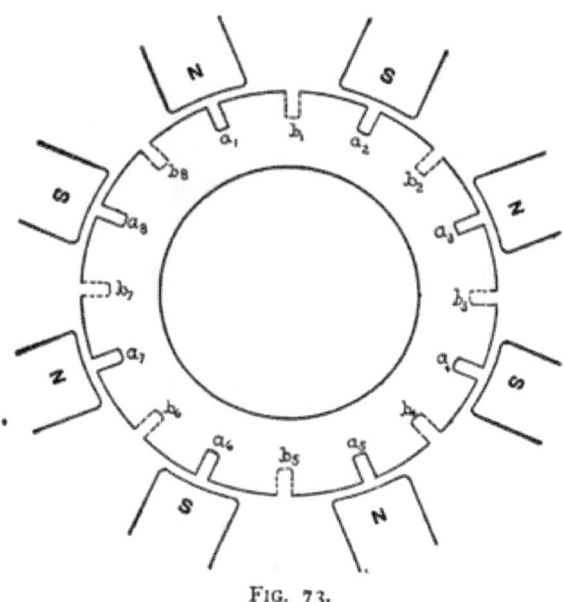

FIG. 73.

such an armature. The slots marked a_1, a_2, a_3, etc., receive the conductors belonging to phase A, and those marked b_1, b_2, b_3, etc., receive those belonging to phase B. The A winding would pass up* slot a_1, down a_2, up a_3 and so on, the terminals of the winding being connected to two collector rings. The B winding would pass up slot b_1, down b_2, up b_3, and so on, its terminals being connected to two collector rings distinct from those to which the A winding is connected.

The armature windings A and B here described are of the concentrated type (see Art. 19) having only one slot per pole for each winding. Distributed windings also are frequently used for two-phase alternators. Thus Fig. 74 shows a portion of a two-phase armature with its A and B windings each distributed in two

---

* Up and down being parallel to the armature shaft *to* and *from* one end of the armature.

SINGLE-PHASE AND POLYPHASE ALTERNATORS. 83

slots per pole. The coils belonging to windings A and B are differently shaded to distinguish them. The coils belonging to

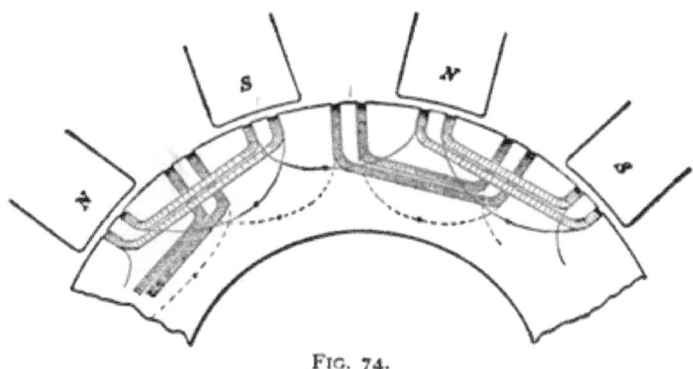

FIG. 74.

winding A are connected together in a manner indicated by the dotted lines and the coils belonging to winding B are connected together as indicated by the full lines. [See Art. 84.]

In general two-phase alternators are provided with two pairs of collector rings; occasionally however one ring is made to serve

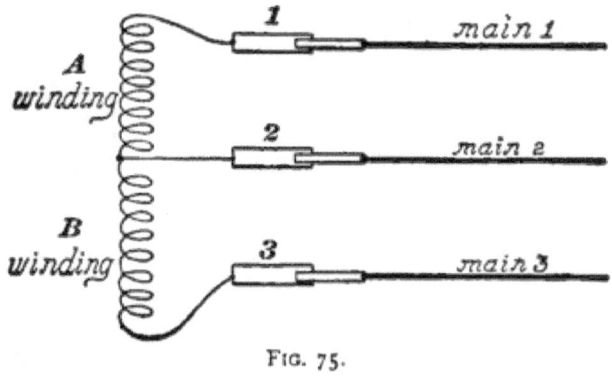

FIG. 75.

as a common connection for the two phases as shown in Fig. 75.

65. Two-phase e. m. f.'s and currents.—The two lines A and B, Fig. 76, represent the e. m. f.'s of the A and B windings respectively of a two-phase alternator. If the circuit which receives current from A is of the same resistance and reactance as the circuit which receives current from *B* then the system is said to be *balanced* and each current lags behind its e. m. f. by the same amount. In this case the currents are equal and in quadrature with each other and are represented by the two dotted lines *a* and *b* in Fig. 76.

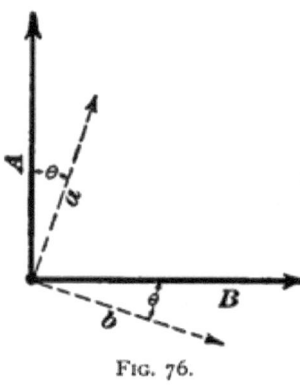

FIG. 76.

66. Electromotive force and current relations in two-phase system. *Electromotive force.*—The e. m. f. between the mains 1 and 3, Fig. 75, is the sum * of the e. m. f.'s *A* and *B*, Fig. 76. This e. m. f. is therefore represented by the diagonal of the parallelogram constructed on *A* and *B*, Fig. 76. It is 45° behind *A* in phase and its effective value is $\sqrt{2}E$ where E is the common effective value of the e. m. f.'s *A* and *B*.

Current.—The current in main 2 is the sum * of the currents in mains 1 and 3, namely *a* and *b*, Fig. 76. This current is therefore represented by the diagonal of the parallelogram constructed on *a* and *b*, Fig. 76; it is 45° behind *a* in phase and its effective value is $\sqrt{2}I$ where I is the common effective value of the currents *a* and *b*.

67. The three-phase alternator.—Consider three similar single-phase armatures, *A*, *B* and *C*, mounted side by side on the same shaft and revolved in the same field. Fix the attention upon a certain armature slot of *A* and let time be reckoned from the instant that this slot is squarely under an *N*-pole. Let *t* be the time which elapses as this armature slot passes from the center of one *N*-pole to the center of the next *N*-pole. The armature *B* is to be so fixed to the shaft that its slots are squarely under the poles at the instant $\frac{1}{3}$ *t*, and the armature *C* is to be so fixed that its slots are squarely under the poles at the instant $\frac{2}{3}$ *t*. While a slot passes from the center of one *N*-pole to the center of the

---

* Or difference according to convention as to sign.

next N-pole the e. m. f. passes through one complete cycle. Hence the e. m. f.'s given by three armatures, arranged as above, will be 120° apart in phase, as shown in Fig. 77, in which the lines A, B and C represent the respective e. m. f.'s. The currents given by the armatures to three similar receiving circuits lag equally behind the respective e. m. f.'s and are represented by the dotted lines a, b and c. This combination of three alternators is called a *three-phase alternator*. In practice the three distinct windings A, B and C are placed upon one and the same armature body. For this purpose the armature body has three times as many slots as A, B or C.

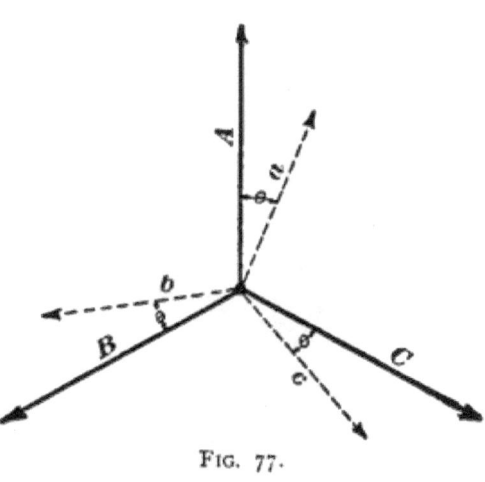

Fig. 77.

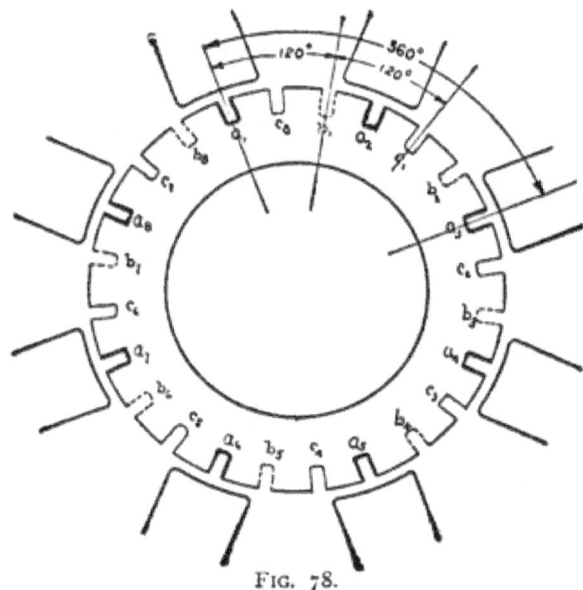

Fig. 78.

Fig. 78 shows the arrangement of the slots for such a winding. The slots belonging to phase A are drawn in heavy lines and are marked a_1, a_2, etc. Those belonging to phase B are shown dotted and those belonging to phase C are shown in light lines. The A winding would pass up slot a_1, down a_2, up a_3, etc.; the B winding, up b_1, down b_2, up b_3, etc.; and similarly for phase C.

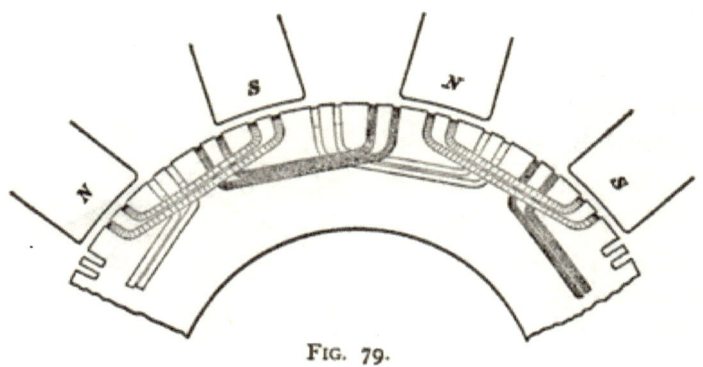

FIG. 79.

The windings A, B and C here described are of the concentrated type, having only one slot per pole for each winding. Distributed windings also are frequently used for three-phase alternators. Thus Fig. 79 shows a portion of a three-phase armature

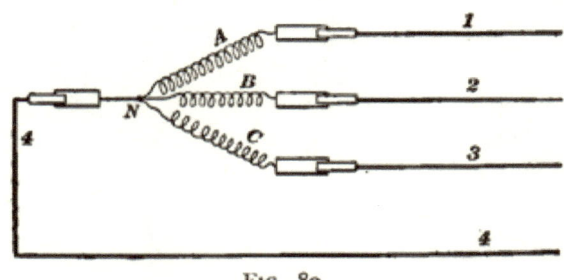

FIG. 80.

with its A, B and C windings each distributed in two slots per pole. The coils belonging to windings A, B and C respectively are differently shaded to distinguish them. The manner of connecting the coils of each winding is described in Art. 84.

If the three circuits of a three-phase alternator are to be entirely independent six collector rings must be used, two for each winding; however, the circuits may be kept practically independent by using four collector rings and four mains, as shown in Fig. 80. The main 4 serves as a common return wire for the independent currents in mains 1, 2 and 3. When the three receiving circuits are equal in resistance and reactance, that is, when the system is balanced, the three currents are equal and 120° apart in phase (each current lagging behind its e. m. f. by the same amount) and their sum is at each instant equal to zero : in

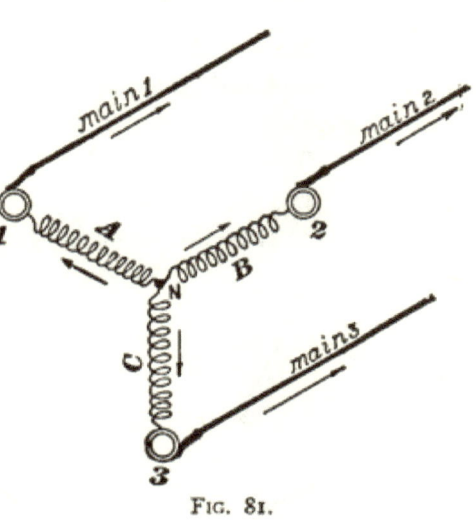

FIG. 81.

which case main 4, Fig. 80, carries no current and this main and the corresponding collector ring may be dispensed with, the three windings being connected together at the point N, called the common junction. This arrangemnnt, shown in the symmetrical diagram, Fig. 81, is called the Y, or star, scheme of connecting the three windings A, B and C.

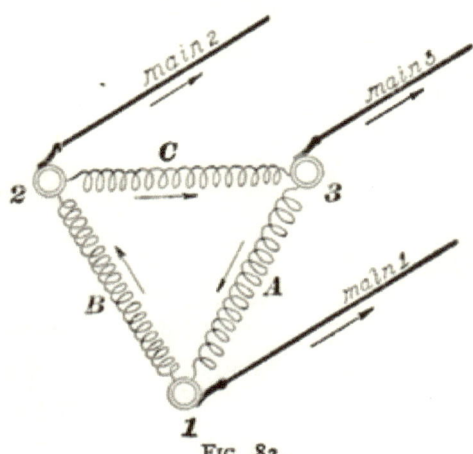

FIG. 82.

Another scheme for connecting the three windings A, B and C (also for balanced loads) called the Δ (delta) or mesh scheme is shown in Fig. 82. Winding A is connected between rings 3 and 1, winding B between rings 1 and 2 and winding C between rings 2 and 3.

The direction in a circuit in which the electromotive force or current is considered as a positive e. m. f. or current is called the *positive direction through the circuit.* This direction is chosen arbitrarily. The arrows in Figs. 81 and 82 indicate the positive directions in the mains and through the windings. It must be remembered that these arrows do not represent the actual directions

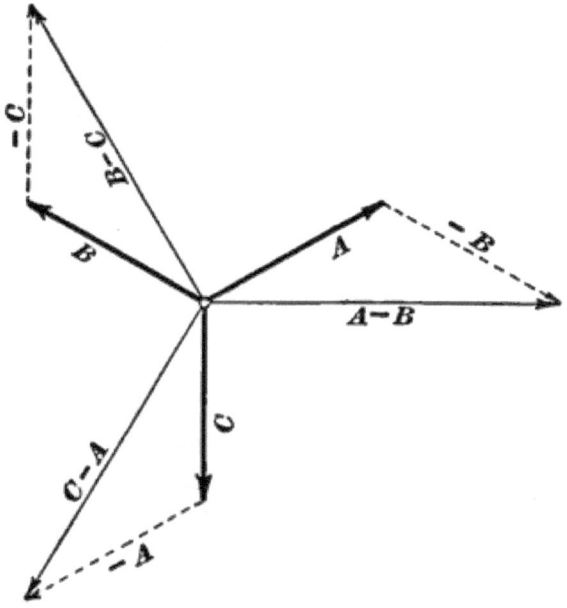

FIG. 83.

of the e. m. f.'s or currents at any given instant but merely the directions of *positive* e. m. f.'s or currents. Thus in Fig. 81 the currents are considered positive when flowing from the common junction towards the collecting rings and the currents are never all of the same sign.

68. Electromotive force and current relations in Y-connected armatures. *E. m. f. relations.*—Passing through the windings A

and B from ring 2 to ring 1,* in Fig. 81 the winding A is passed through in the positive direction and the winding B in the negative direction. Therefore, the e. m. f. between mains 1 and 2 is $A - B$ (see Fig. 77). Similarly the e. m. f. between mains 2 and 3 is $B - C$ and the e. m. f. between mains 3 and 1 is $C - A$. These differences are shown in Fig. 83. The e. m. f. between mains 1 and 2, namely, $A - B$, is 30° behind A in phase and its effective value is $2E \cos 30° = \sqrt{3}E$, where E is the common value of each of the e. m. f.'s, A, B and C. Similar statements hold concerning the e. m. f.'s between mains 2 and 3 and between mains 3 and 1. Hence the e. m. f. between any pair of mains leading from a three-phase alternator with a Y-connected armature is equal to the e. m. f. generated per phase multiplied by $\sqrt{3}$.

Current relations.— In the Y connection the currents in the mains are equal to the currents in the respective windings, as is evident from Fig. 81.

69. Electromotive force and current relations in Δ connected armatures. *E. m. f. relations.*—In Δ connected armatures the e. m. f.'s between the mains or collector rings are equal

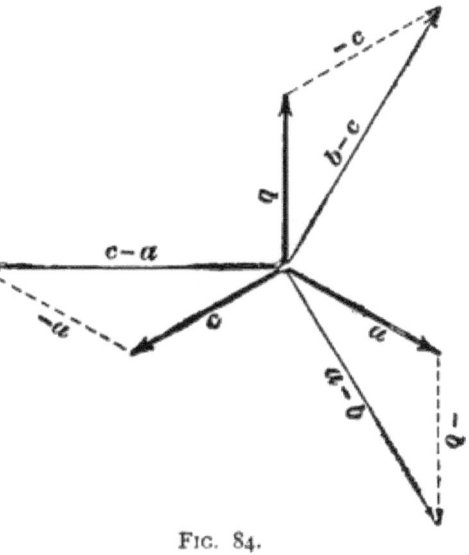

FIG. 84.

to the e. m. f.'s of the respective windings as is evident from Fig. 82.

Current relations.—Referring to Fig. 82 we see that a positive current in winding A produces a positive current in main 1 and

---

* Which is the direction in which an e. m. f. must be generated to give an e. m. f., acting upon a receiving circuit from main 1 to main 2.

that a negative current in winding B produces a positive current in main 1, therefore the current in main 1 is $a-b$ when a is the current in A and b is the current in B. Similarly the current in main 2 is $b-c$ and the current in main 3 is $c-a$. These differences are shown in Fig. 84. The current in 1, namely $a-b$, is 30° behind a in phase and its effective value is $\sqrt{3}\, I$ when I is the common effective value of the currents a, b, c in the different phases. Similar statements hold for the currents in mains 2 and 3; so that the current in each main of a Δ connected armature is $\sqrt{3}$ times the current in each winding.

70. Connection of receiving circuits to three-phase mains. *In case of dissimilar circuits* (unbalanced system).—When the receiving circuits which take current from three-phase mains are dissimilar, four mains should be employed as indicated in Fig. 80; each receiving circuit being connected from main 4 to one of the other mains. It is however desirable to keep the three windings A, B and C of the alternator as nearly equally loaded as possible, and the receiving circuits are so disposed in practice as to satisfy this condition as nearly as possible.

In case of similar circuits (balanced system).—When three-phase currents are used to drive induction motors, synchronous motors or rotary converters, each unit takes current equally from the three mains, and since three-phase currents are utilized mainly in the operation of the machines mentioned, the system is usually balanced. In this case three mains only are employed and each receiving unit has three similar receiving

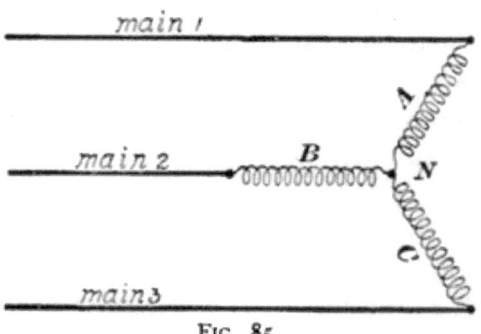

Fig. 85.

circuits connected to the mains according to either the Y or Δ method. The Y method of connecting receiving circuits is shown in Fig. 85. One terminal of each receiving circuit is connected to a main and the other terminals are connected together at N. In this case the current in each receiving circuit is equal to the current in the main to which it is connected. The e. m. f. between the terminals of each receiving circuit is equal to $\dfrac{E}{\sqrt{3}}$ where E is the e. m. f. between any pair of mains.

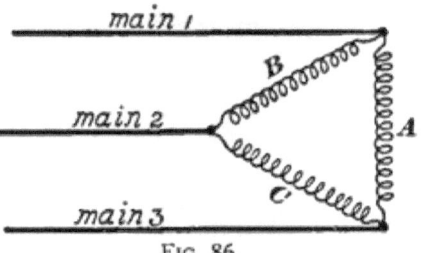

FIG. 86.

The ⌐ method of connecting receiving circuits is shown in Fig. 86. Here the three receiving circuits are connected between the respective pairs of mains, the e. m. f. acting on each receiving circuit is the e. m. f. between the mains, and the current in each receiving circuit is $\dfrac{I}{\sqrt{3}}$ where I is the current in each main.

71. Power in polyphase systems.—The several circuits of a polyphase system are in general entirely separate and independent, and the total power delivered to a receiving apparatus is to be found by measuring the power delivered to each separate receiving circuit; the total power delivered is the sum of the amounts delivered to the different receiving circuits.

Balanced systems.—When a polyphase system is balanced the several circuits are entirely similar and the same amount of power is delivered to each receiving circuit of a given piece of receiving apparatus.

Balanced two phase.—Let E be the e. m. f. of each phase, I the current furnished to each of two similar receiving circuits, and $\cos\theta$ the power factor of each receiving circuit. Then $EI\cos\theta$ is the power delivered to each circuit so that the total power delivered is

$$P = 2EI\cos\theta. \tag{55}$$

Balanced three phase.—Let E be the e. m. f. between the terminals of each receiving circuit, I the current in each receiving circuit, and $\cos\theta$ the power factor of each circuit. Then $EI\cos\theta$ is the power delivered to each receiving circuit so that the total power delivered is

$$P = 3\,EI\cos\theta \tag{56}$$

in which, as must be remembered, E is the e. m. f. at the terminals of each receiving circuit and I is the current in each receiving circuit. On the other hand:

$$P = \sqrt{3}\,EI\cos\theta \tag{57}$$

in which E is the e. m. f. between each pair of mains, I is the current in each main, and $\cos\theta$ is the power factor of each receiving circuit. Equation (57) may be derived from (56) by considering that the current I_m in each main, for the case of Δ connection for example, is equal to $\sqrt{3}\,I$, so that, substituting $\dfrac{I_m}{\sqrt{3}}$ for I in equation (56) we have equation (57).

72. The flow of energy in balanced polyphase systems.—It was pointed out in Art. 22 that the power developed by a single-phase alternator pulsates with the alternations of e. m. f. and current. The power delivered to a balanced system by a polyphase generator, on the other hand, is not subject to pulsations but is entirely steady and constant in value.

Discussion for a two-phase alternator: Consider a single-phase alternator of which the e. m. f. is:

$$e = E\sin\omega t \tag{a}$$

and which gives a current

$$i = I\sin(\omega t - \theta)$$

or
$$i = I\sin\omega t \cos\theta - I\cos\omega t \sin\theta. \tag{b}$$

The instantaneous power ei is:

$$ei = EI\cos\theta \sin^2\omega t - EI\sin\theta \sin\omega t \cos\omega t$$

which pulsates with a frequency twice as great as the frequency of e and i.

Let equations (a) and (b) express the e. m. f. and current of one phase of a (balanced) two-phase alternator, then the e. m. f. and current of the other phase are

$$e' = E\cos\omega t \tag{c}$$
$$i' = I\cos(\omega t - \theta) = I\cos\omega t \cos\theta + I\sin\omega t \sin\theta. \tag{d}$$

The instantaneous power output of this phase is

$$e'i' = EI \cos \theta \cos^2 \omega t + EI \sin \theta \sin \omega t \cos \omega t.$$

Therefore the total power output of the two-phase machine is

$$ei + e'i' = EI \cos \theta (\sin^2 \omega t + \cos^2 \omega t)$$
$$= EI \cos \theta$$

which is constant.

Remark: The torque of a single-phase alternator pulsates with the pulsations of the power output. In a balanced polyphase alternator however the torque is steady since the power does not pulsate; also polyphase synchronous motors, rotary converters, and induction motors are driven by a steady torque.

73. Measurement of power in polyphase systems.—In a polyphase system, balanced or unbalanced, the power taken by any unit, such as an induction motor, may be determined by measuring the power taken by each single receiving circuit and adding the results. In order to measure the power taken by a single receiving circuit the current coil of the wattmeter is connected in series with the circuit and the fine wire coil is connected to the terminals of the circuit. The inconvenience of connecting and

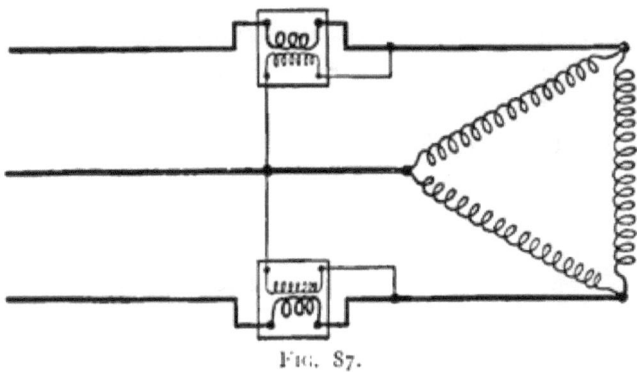

Fig. 87.

disconnecting the wattmeters makes it necessary to use a separate wattmeter for each receiving circuit. Two wattmeters are sufficient for the complete measurements of the power taken by any three-phase receiving unit. The connections are shown in Fig.

87. The receiving circuits may be balanced or unbalanced and connected Y or Δ.

In a balanced polyphase system, a condition which is seldom strictly realized, the power taken by one only, of the receiving circuits need be measured.

Problems.

1. A common return wire is used for the two currents of a two-phase system. The system is balanced and each current is equal to 100 amperes. What is the current in the common return wire?

2. The e. m. f. of each phase, problem 1, is 500 volts. What is the e. m. f. between the outside wires?

3. Three similar receiving circuits are Δ connected to 3-phase mains, the e. m. f. between each pair of mains being 110 volts. The power delivered to the three circuits is 150 kilowatts and the power factor of each circuit is .90. What is the current in each circuit and in each main?

4. Three similar receiving circuits are Y connected to the 3-phase mains, problem 3; the total power delivered is 150 kilowatts and the power factor of each circuit is .90. What is the current in each circuit and in each main; and what is the e. m. f. between the terminals of each circuit?

CHAPTER IX.

ALTERNATORS.

(Continued.)

74. Armature reaction.—The amount of magnetic flux entering the armature core from the field poles and the manner of its distribution over the pole faces, depend upon the combined magnetic action of the field current and of the armature current.

Distortion of field.—The armature current in an alternator tends to concentrate the magnetic flux under the trailing horns of the pole pieces very much as in the direct current dynamo. The effect of this concentration of flux is to slightly increase the magnetic reluctance of the air gap and of the saturated portions of the pole pieces and armature core. This increase of magnetic reluctance causes a decrease of flux and a consequent decrease of the e. m. f. of the alternator, others things being equal. This effect may be appreciable, but it cannot be accurately calculated.

Magnetizing and demagnetizing action of the armature current.—When the current given by an alternator is in phase with its e. m. f. the only effect of the armature current upon the field is the distorting effect described above. When the current is not in phase with the e. m. f. the distorting effect is decreased* and in addition there is a magnetizing action or demagnetizing action upon the field according as the current is *ahead* of or *behind* the e. m. f. in phase.

Consider a bundle of Z armature wires, grouped in a slot, for example. Let

$$e = E \sin \omega t.$$

be the alternating e. m. f. induced in this bundle of conductors. This e. m. f. is a maximum when the slot is at a, Fig. 88. It is zero when the slot is at b

---

* The distorting effect is due to the component of the current parallel to the e. m. f. and the magnetizing effect is due to the component at right angles to the e. m. f.

and it is a minimum (negative maximum) at c. Therefore the value of ωt is 90° at a, 180° at b and 270° at c. Let i be a given current flowing in the bundle of wires. The *ampere-turns* of the bundle is then Zi. If the bundle of wires is at a its ampere-turns will be without appreciable effect on the magnetic circuit $m\ m\ m$ shown by the dotted line; at b the ampere-turns will have their full demagnetizing effect (negative)* upon the magnetic circuit, and at c the effect will again be zero.

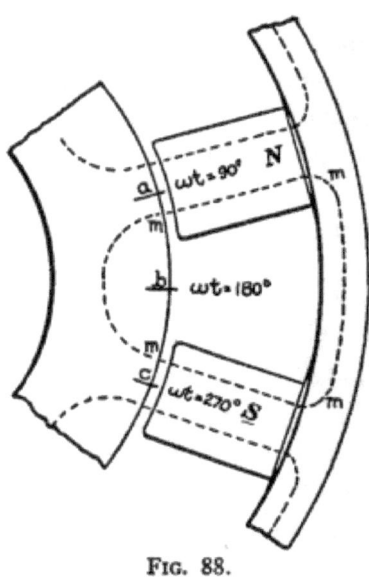

Fig. 88.

Now, $\cos \omega t$ is zero at a, negative unity at b and zero at c. Therefore $Zi \cos \omega t$ is an expression which gives the true magnetic effect of the bundle of wires at a, b and c. *We assume his expression to hold for all points between a and c.* The actual current in the bundle of wires is

$$i = I \sin (\omega t - \theta)$$

in which θ is the angle of lag of the current behind the e. m. f. Therefore substituting this value of i in the expression $Zi \cos \omega t$ we have

$$m = ZI \cos \omega t \sin (\omega t - \theta).$$

in which m is the effective ampere-turns of the bundle of wires at the instant t. To find the average value of m expand $\sin (\omega t - \theta)$, whence

$$m = ZI \sin \omega t \cos \omega t \cos \theta - ZI \cos^2 \omega t \sin \theta.$$

Now the average value of m is the sum of the average values of the two terms of the right hand member; but the average value of $\sin \omega t \cos \omega t$ is zero and the average value of $\cos^2 \omega t$ is $\tfrac{1}{2}$. Therefore

$$\text{average } m = - \tfrac{1}{2} ZI \sin \theta$$

or putting $\sqrt{2}\, I$ for I we have

$$\text{average magnetizing ampere-turns} = -.707\ ZI \sin \theta. \qquad (58)$$

When this formula is applied to an alternator the windings of which are not too widely distributed, Z is to be taken as the total number of armature conductors divided by the number of poles. This equation shows that when the current lags behind the e. m. f. (angle θ positive) the armature current weakens the field and *vice versa*.

75. Armature inductance.—The effect of the inductance of the armature of an alternator is to cause the e. m. f. between the col-

---

* The current i is considered positive when it is in the direction of the e. m. f. which is induced under the N pole. A current in this direction has a demagnetizing effect for all positions of the slot between a and c.

lecting rings to fall off considerably with increasing current, especially if the receiving circuit is inductive, as is explained in the following article.

The inductance of an alternator armature is especially great if the conductors are deeply imbedded in the armature core, and the inductance is increased by the grouping of the conductors together in bundles. This is evident from equation (7) which expresses the inductance of a bundle of Z wires surrounded by iron. This equation (7) shows that inductance is proportional to the square of the number of turns.

76. The electromotive force lost in the armature. *Armature drop.*—The e. m. f. at the collecting rings of an alternator is less than the total e. m. f. induced in the armature for the reason that a portion of the induced e. m. f. is used to overcome the resistance and a portion is also used to overcome the inductance of the armature windings.

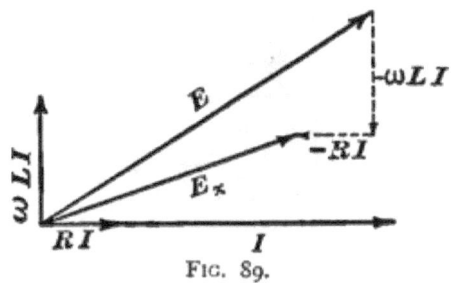

FIG. 89.

General case.—Let I, Fig. 89, be the current given by the alternator and E the total induced e. m. f. Then $\omega L I$ is the portion of E used to overcome the inductance L of the armature, and RI is the portion of E used to overcome the resistance of R of the armature. Subtracting $\omega L I$ and RI from E gives the e. m. f. at the collecting rings, or "external e. m. f.," E_x.

Armature drop, non-inductive load.—In this case E is nearly in phase with I and the subtraction of $\omega L I$ from E scarcely reduces its value, $\omega L I$ being nearly at right angles to E. Therefore with a non-inductive receiving circuit the armature drop depends almost wholly upon the armature resistance.

Armature drop, inductive load.—When the phase difference between E and I is nearly 90°, then the subtraction of RI from E

scarcely reduces its value. Therefore with a highly inductive receiving circuit the armature drop depends almost wholly upon the armature inductance.

77. The characteristic curve of the alternator.—The curve obtained by plotting observed values of the external e. m. f. for various currents taken from an alternator, is called the characteristic curve of the alternator. Such characteristic curves are shown in Fig. 90. Curve A is for a separately excited alternator having but small armature inductance and curve B is for a separately excited alternator having large armature inductance. The shape of the characteristic curve of a given alternator depends to a greater or less extent upon the inductance of the receiving circuit. The falling off of e. m. f. with increase of current is due in part to the demagnetizing action of the armature current which weakens the field, and in part to the increased armature drop with increase of current.

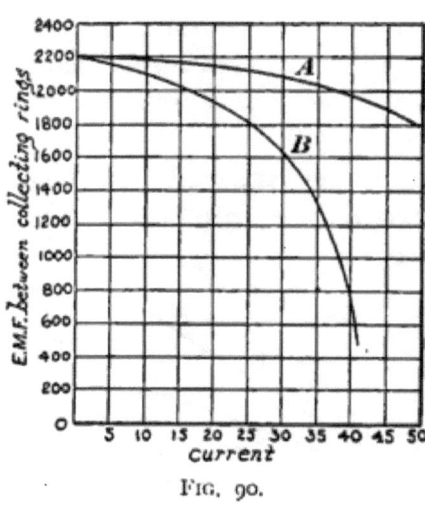

FIG. 90.

78. The constant current alternator.—An alternator of which the armature has an excessive inductacne, or an ordinary alternator in circuit with which a large inductance is connected, gives a current which does not vary greatly with the resistance* of the receiving circuit.

This may be shown as follows: Let E, Fig. 91, be the total induced e. m. f. of an alternator sending

---

* Unless the resistance becomes very large

ALTERNATORS. 99

current through a circuit of which the reactance ωL is constant and large, compared with the resistance R. Then ωLI will be large, compared with RI. Further, RI and ωLI are at right angles to each other and their vector sum is E, so that the point P, Fig. 91, lies on a semicircle constructed on E as a diameter. Now, when RI is small, compared with E, then ωLI is very nearly equal to E, that is, ωLI is approximately constant and, therefore, I is approximately constant.

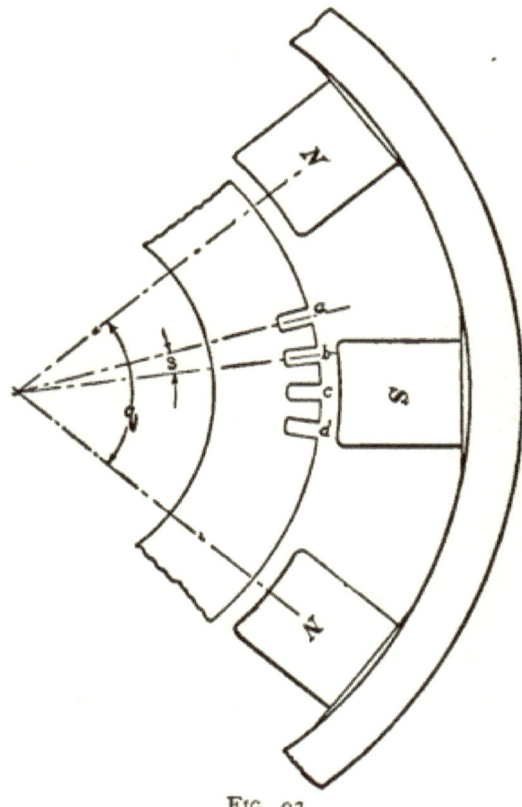

FIG. 92.

79. Effect of distributed winding upon the e. m. f. of an alternator.*—Consider an armature winding, A, concentrated in a set

---

* This question is discussed in a slightly different manner in the chapter on the rotary converter.

of slots, one slot per pole. The effective e. m. f. of this winding is

$$A = \frac{4.44\ NTf}{10^8}$$

according to equation (22), Chap. II. Suppose another similar concentrated winding, B, is placed upon the same armature in slots distant s from the first set of slots, s being the angle shown in Fig. 92. This figure shows one slot only of the first set and one slot only of the second set. The phase difference between the e. m. f.'s in these two windings is the angle $\frac{s}{q} \times 360°$, inasmuch as the angle q from N to N is equivalent to $360°$ of phase difference. These two e. m. f.'s are represented by the lines A

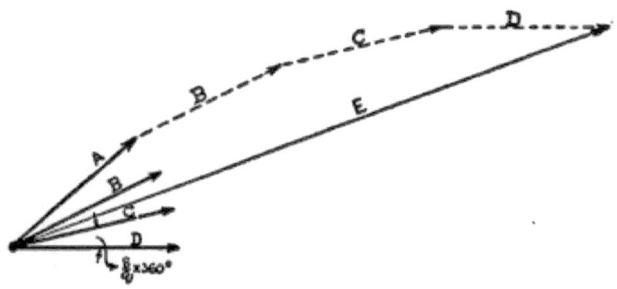

FIG. 93.

and B, Fig. 93. Similarly the lines C and D represent the e. m. f.'s in two additional similar windings concentrated in two additional sets of slots c and d, Fig. 92. If all these windings are connected in series the effective e. m. f. produced will be the vector sum E of A, B, C and D. If we were to calculate the effective e. m. f. produced by A, B, C and D in series on the assumption that all the windings are concentrated in one set of slots, that is, if we were to calculate this total e. m. f. by means of equation (22), using for T the total number of turns in all the windings, we would get a result greater than E in the ratio of the sum of the sides A, B, C and D of the polygon to the chord E, Fig. 93. This ratio may be called the *phase constant* of the distributed winding. By introducing the phase constant k in equation (22) this equation becomes

$$E = \frac{4.44kNTf}{10^8} \qquad (59)$$

This form of the fundamental equation of the alternator is applicable to armatures with distributed windings. The following table gives the values of k for various degrees of distribution. The slots for a given winding are always grouped so many per pole and a group of slots may cover $\frac{1}{4}$, $\frac{1}{3}$, $\frac{1}{2}$, etc., of the

VALUES OF PHASE CONSTANT k FOR DISTRIBUTED WINDINGS.

Number of slots in each Group.	Widths of groups of slots in fractional parts of N to S.				
	¼	⅓	½	¾	Whole.
1	1.000	1.000	1.000	1.000	1.000
2	.980	.966	.924	.831	.707
3	.977	.960	.912	.805	.666
4	.976	.958	.908	.795	.653
Infinity.	.975	.955	.901	.784	.637

Note: Column headed ⅓ applies to 3-phase alternators.
Column headed ½ applies to 2-phase alternators.
Width of group $= ns$, where n is number of slots in a group and s is distance from center to center of adjacent slots.

space from the center of an N pole to the center of an s pole. Thus in Fig. 94 is shown an 8-pole machine of which the armature is slotted for a distributed winding, there being three slots per pole, these slots being grouped so as to cover ⅓ of the space from an N to an S pole. In the table the width of a group of slots is ns where n is the number of slots in a group and s is the distance from center to center of adjacent slots.

80. Practical and ultimate limits of output.—The dotted curve, Fig. 95, is the characteristic curve of a given alternator. This curve shows the relation between the current output and the e. m. f. between the collecting rings, the field excitation being kept constant. The ordinates of the full line curve represent the power outputs corresponding to the different currents (receiving

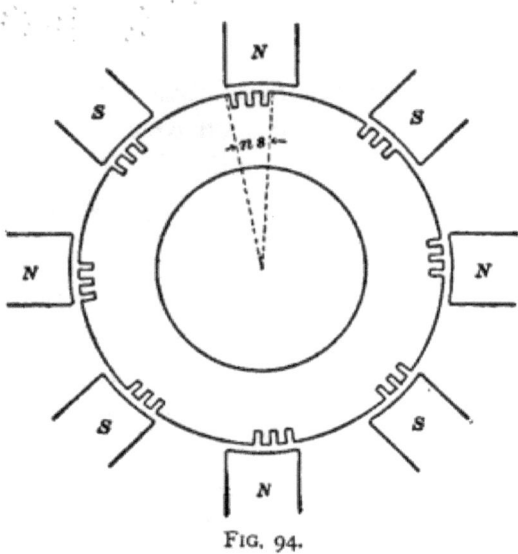

FIG. 94.

circuit non-inductive). The maximum output of the alternator is thus 68 kilowatts when the current output is 38 amperes. In practice the allowable power output of an alternator is limited to a smaller value than this maximum output by one or the other of the following considerations.

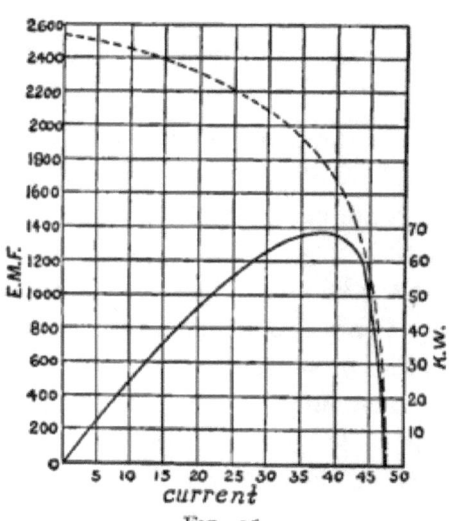

FIG. 95.

(a) Electric lighting and power service usually demands an approximately constant c. m. f. and it is not permissible to take from an alternator so large a current as to greatly reduce its e. m. f. This difficulty may be largely overcome by providing for an increase of field excitation of the alternator with increase of load as is done in the alternator with a compound field winding. See Art. 89.

(*b*) The current delivered by an alternator generates heat* in the armature of the alternator and the temperature of the armature rises until it radiates heat as fast as heat is generated in it by the current. Excessive heating of the armature endangers the insulation of the windings and it is not permissible to take from an alternator so large a current as to heat its armature more than 40° or 50° C. above the temperature of the surrounding air. This heating effect of the armature currents usually fixes the allowable output of an alternator except in those rare cases where extreme steadiness of e. m. f. is required, or where the alternator is not compounded.

Influence of inductance upon output.—An alternator is rated according to the power it can deliver steadily to a non-inductive receiving circuit without overheating. The amount of power which an alternator can satisfactorily deliver to an inductive receiving circuit is less than that which it can deliver to a non-inductive receiving circuit, because of the phase difference of e. m. f. and current. The cosine of this phase difference ($\cos \theta$) is called the power factor of the receiving circuit as before pointed out. The power factor of lighting circuits is very nearly unity.

The power factor of induction motors, synchronous motors and rotary converters is often as low as .75 and sometimes even less.

81. Frequencies.—The frequencies employed in practice range from 20 or 25 to 150 cycles per second. Very low frequencies are not suitable for lighting on account of the tendency to produce flickering of the lights; on the other hand high frequencies, which tend to make transformers cheaper for a given output, are entirely satisfactory and are often employed for lighting.

High frequencies are not well adapted for the operation of in-

---

* Additional heat is generated in the armature by the hysteresis and eddy current losses in the armature core.

duction motors, synchronous motors and rotary converters because high frequencies necessitate either great speed or a great number of poles. For such purposes frequencies as low as 25 per second are often employed.

A frequency of 60 has been quite generally adopted for machines used to operate both lights and motors.

82. Speeds. Number of poles.—A machine which is to be belt-driven may be driven as fast as is compatible with the strength and rigidity of the rotating part. The allowable speed of rotation in ordinary dynamos and alternators is such as will give a peripheral velocity of from 4000 to 6000 feet per minute. When a machine is direct-connected to an engine or water wheel its speed is fixed by that of the prime mover.

The number of poles depends upon the speed of an alternator and the frequency it is to give, according to equation (18). Large machines as a rule must run slower than small ones and they, therefore, have a greater number of poles. The accompanying table gives data as to speed, frequency, and number of poles of a few recent American machines. Machines 7 and 8 are of the direct-connected type.

TABLE.

	125 Cycles.				60 Cycles.		
No.	No of poles.	Output K. W.	Speed r. p. m.	No.	No. of poles.	Output K. W.	Speed r. p. m.
1	10	60	1500	4	8	75	900
2	14	125	1070	5	12	150	600
3	16	200	937	6	16	250	450
				7	36	250	200
				8	40	750	180

83. Armatures.—Alternator armatures are usually of the drum type or disc type. The former type is almost universal in America while the disc type is frequently used in England; for example, the Ferranti and Mordey machines have disc armatures

Drum armatures have laminated iron cores similar to the armature cores used for direct-current dynamos, while disc armatures are usually made up without iron. Ring armatures have been used only to a very limited extent.

Drum armatures are, in nearly all modern machines, of the toothed or ironclad type. The conductors are bedded in slots. This has the double advantage of shortening the gap space from pole face to armature core and of protecting the armature conductors from injury. One type of such an armature has already been shown in Fig. 10, the heavy coils being first wound on forms and then pressed into position on the armature core. When distributed windings are used straight slots as shown in Fig. 96 are

FIG. 96.   FIG. 97.

often employed. Fig. 97 shows a style of slot commonly used in which the coils are held in position by the wooden wedge W. Armature core discs should be varnished, japanned or in some manner insulated from each other to prevent eddy currents. This is especially necessary in the case of alternator armature cores because the frequency is comparatively high.

84. Armature windings.—Any direct current dynamo* may be converted into a single-phase or polyphase alternator by providing it with collecting rings as explained in the chapter on the rotary converter. Ordinarily, however, the armature windings of alternators are very different from the armature windings of direct-current dynamos. In the type of winding most frequently employed a number of distinct coils are arranged on the armature; in these coils alternating e. m. f.'s are induced as they pass

---

* Except the so-called unipolar dynamo.

the field magnet poles, and these coils are connected in series between the collecting rings if high e. m. f. is desired or in parallel* between the collecting rings if low e. m. f. is desired.

Single-phase winding.—Fig. 10 shows a common type of single-phase winding having one coil per pole. Fig. 98 shows another

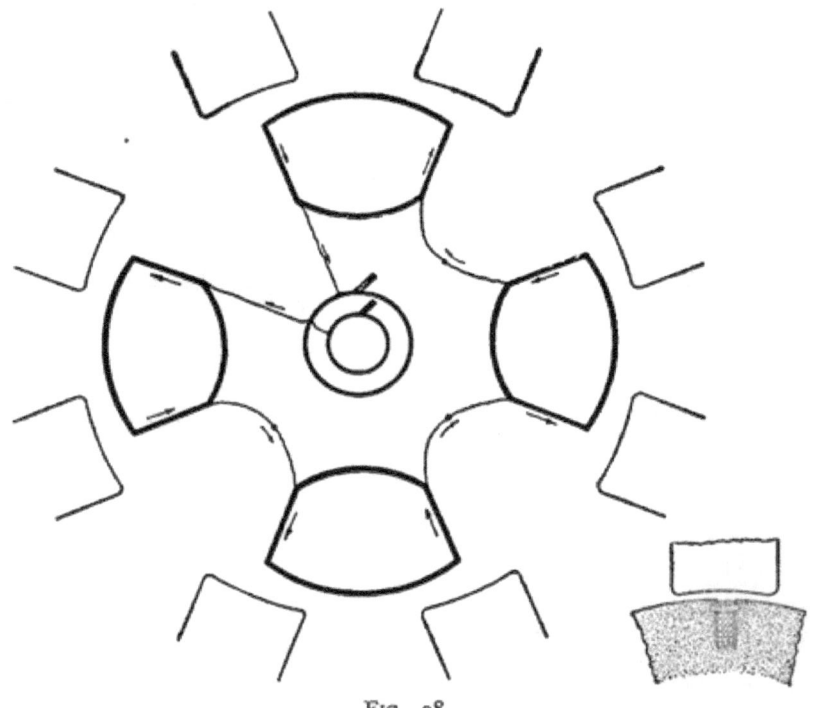

FIG. 98.

type of concentrated single-phase winding having one coil to each pair of poles or one slot per pole. In the diagram, Fig. 98, the heavy sector shaped figures represent the coils and the light lines represent the connections between the terminals of the coils. The radial parts of the sector shaped figures represent the portions of the coils which lie in the slots, and the curved parts rep-

---

* The coils of a distributed winding cannot all be connected in parallel between the collecting rings for the reason that the induced e. m. f.'s in the various coils are not exactly in phase and local currents would circulate in the coils if connected in parallel.

resent the ends of the coils. The circles at the center of the figure represent the collecting rings, one being shown inside the other for clearness. The arrows represent the direction of the current at a given instant. All e. m. f.'s under N poles are in one direction and all e. m. f.'s induced under S poles are in the opposite direction. These remarks apply to Figs. 98 to 105 inclusive. Fig. 99 represents a single-phase winding distributed in

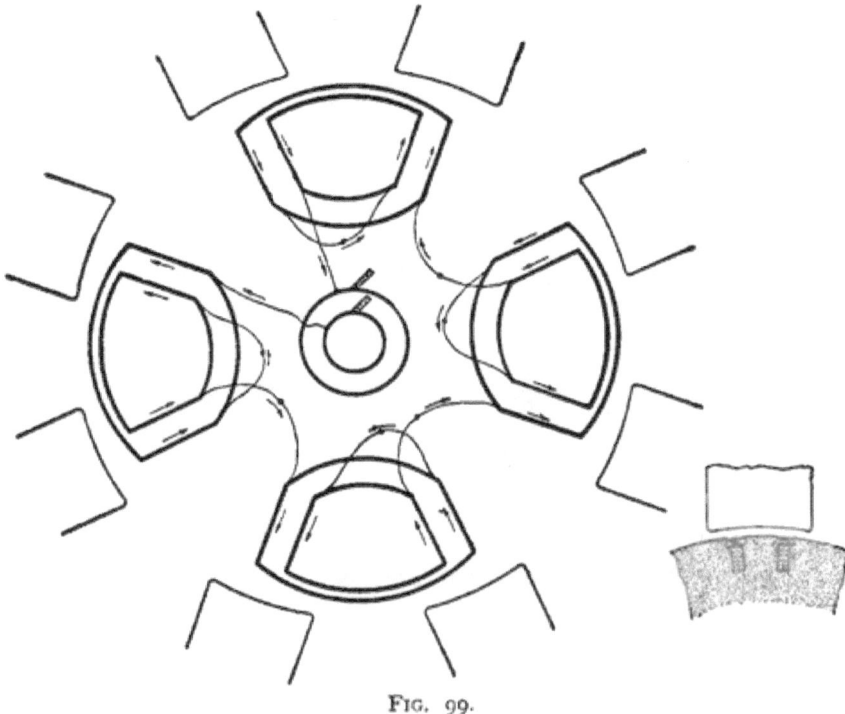

FIG. 99.

two slots per pole, all the coils being connected in series. Fig. 99 is a type of winding which, for the same number of conductors, has a smaller inductance than the type shown in Fig. 98 and the armature shown in Fig. 99, for the same number of conductors, gives a smaller e. m. f. than the armature shown in Fig. 98.

Two-phase windings.—The two-phase winding is two independent single-phase windings on the same armature, each being connected to a separate pair of collecting rings, as shown in Figs. 100 and 101. Fig. 100 shows a two-phase concentrated winding, one slot per pole for each phase. Fig. 101 shows a two-

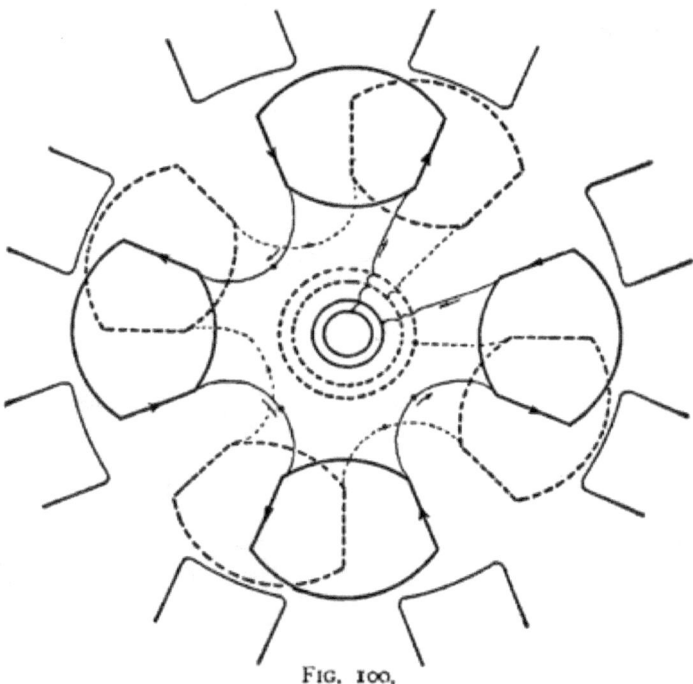

FIG. 100.

phase winding distributed in two slots per pole for each phase.

Three-phase windings.—The three-phase winding is three independent single-phase windings on the same armature, the terminals of the individual windings being connected according to the Y-scheme or Δ-scheme, as explained in Art. 67. Fig. 102 shows a three-phase concentrated winding, one slot per pole for each phase; Y connected. Fig. 103 shows the same winding Δ connected. The Y connection gives $\sqrt{3}$ times as much e. m. f. between collecting rings as the Δ connection for the same winding. The Y connection is more suitable for high e. m. f. machines and the Δ connection for machines for large current output. The line current is $\sqrt{3}$ times as great as the current in each winding

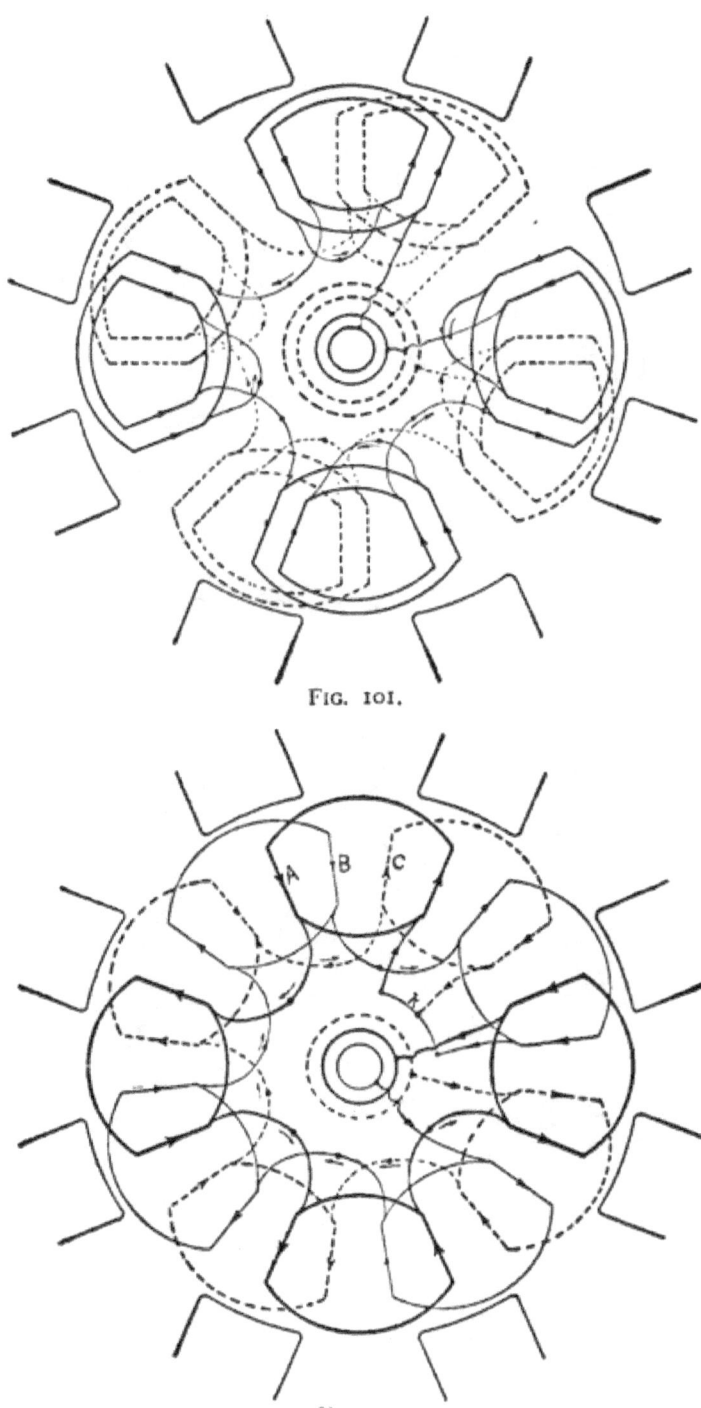

Fig. 101.

Fig. 102.

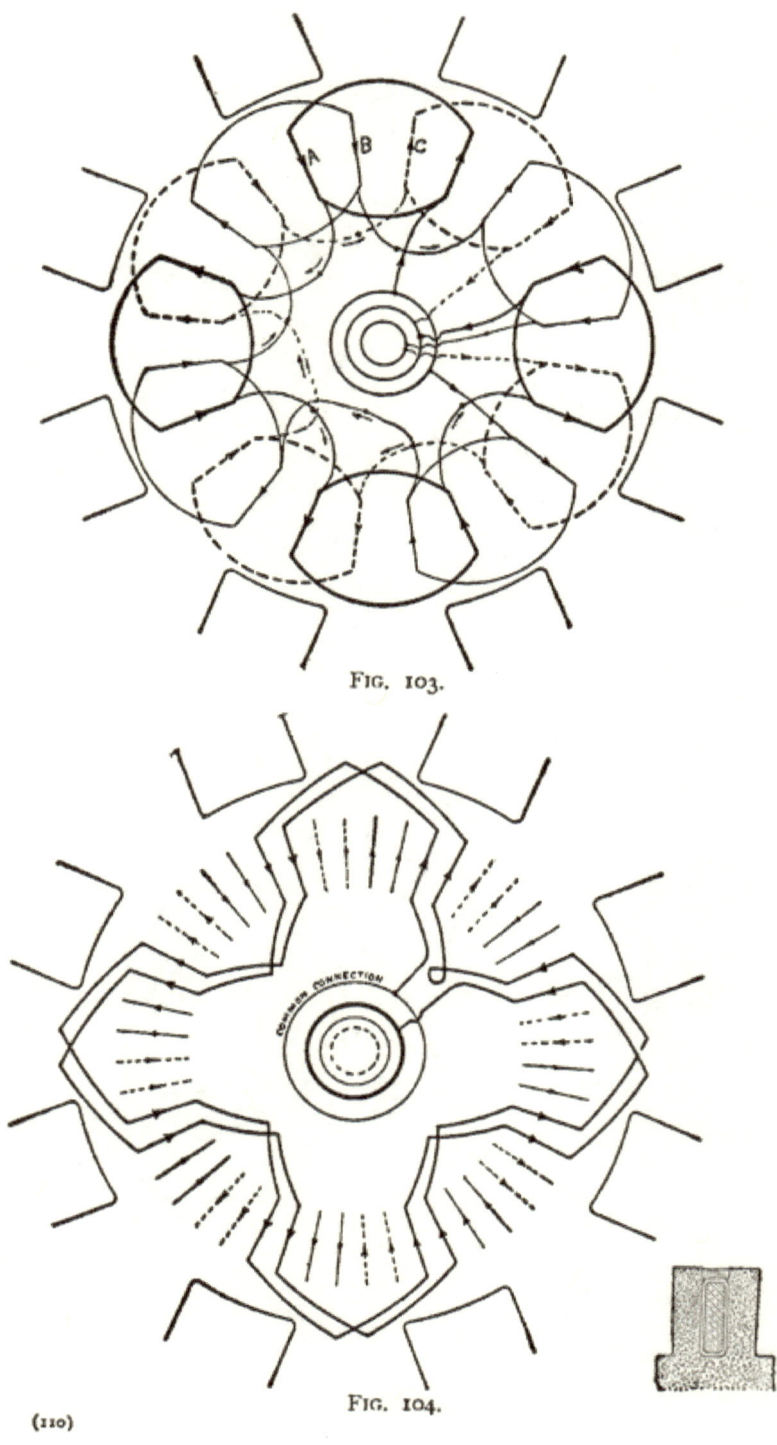

Fig. 103.

Fig. 104.

in a Δ connected armature. Fig. 104 shows a three-phase bar* winding distributed in two slots per pole for each phase. Fig. 105 shows a three-phase coil winding distributed in two slots per pole for each phase and arranged in two layers, there being as many coils on the armature as there are slots, so that portions of

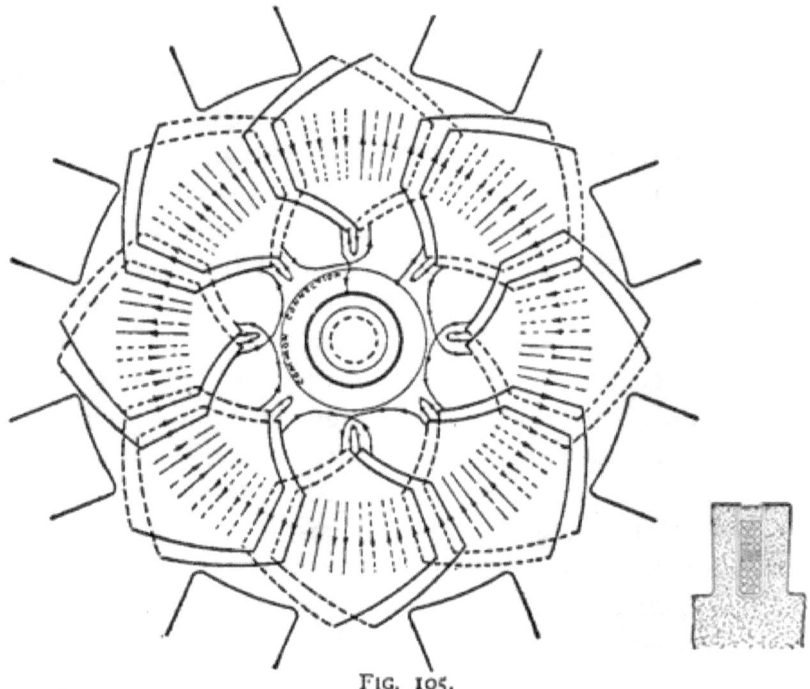

FIG. 105.

two coils lie in each slot, one above the other. The portions of the coils represented by full lines lie in the upper parts of the slots and the adjacent dotted portions lie in the bottoms of the same slots.

The Y connection.—The terminals of the individual windings which are to be connected to the common junction and to the collecting rings may be determined as follows: Consider the instant when winding A is squarely under the pole as shown in Fig. 102; the e. m. f. in this winding (and current also if the circuit is non-inductive) is a maximum and the currents in the other two phases B and C are half as

* One conductor in each slot. This conductor is usually in the form of a copper bar of rectangular cross section.

great. If winding A is connected so that its current is flowing away from K, windings B and C must be connected so that their currents flow towards K.

The $\triangle$ connection.—The three windings form a closed circuit when $\triangle$ connected. The total e. m. f. around this circuit at any instant must be zero. Therefore the e. m. f. in winding A when it is directly under the poles must oppose the e. m. f.'s of windings B and C.

85. Insulation of armatures.—Armatures for alternators must be well insulated in cases where the e. m. f. generated is high as the e. m. f., tending to break down the insulation is the maximum value of the e. m. f. generated, and this is considerably greater than the rated or effective e. m. f. Concentrated or partially distributed windings admit of quite a high degree of insulation inasmuch as the slots may be made quite large and there are comparatively few crossings of the coils at the ends of the armature. Distributed windings can not be so highly insulated because there are many crossings of the coils and the slots are necessarily small, such windings are, therefore, not suitable for the generation of high e. m. f.'s. Alternators having this type of winding should therefore be used in connection with step-up transformers if a high e. m. f. is desired. When it is desired to generate a high pressure directly it is best to use a machine with a stationary armature. Such armatures have been built for e. m. f.'s of 8,000 or 10,000 volts, thus doing away with the necessity of step-up transformers for power transmission lines of moderate length. There is usually more room for thorough insulation on such armatures and the insulation is less liable to deteriorate as it is not disturbed in any way by motion of the armature. Moreover, the use of the stationary armature does away with collector rings and brushes (for the armature) and the consequent necessity of their insulation for high potentials.

The individual coils of an alternator armature are generally heavily taped and treated with insulating oil or varnish, the slots are lined with heavy tubes built up of paper and mica and all parts of the core which are near the coils are also covered with a heavy layer of insulating material.

85. Magnetic densities in armature and air gap.—The armature core is usually made of sufficient cross section to insure a fairly low magnetic density. This is done in order to keep down the hysteresis and eddy current losses which would otherwise be high on account of the comparatively high frequencies employed. The allowable magnetic density in the armature core depends largely upon the frequency, since the density for a given loss may be higher the lower the frequency. The following table from Kolben gives values of the density suitable for various frequencies.

B. Lines per cm².

40 cycles	6500 to	5500
50 "	6000 "	5000
60 "	5000 "	4500
80 "	4500 "	4000
100 "	4000 "	3500
120 "	3500 "	3000

The allowable magnetic density in the air gap will depend to some extent upon the material used for the pole pieces. With cast-iron poles this density should not exceed 4000 to 4500 lines per sq. cm.; with wrought-iron pole pieces it may be as high as 6000 to 7000 lines per sq. cm.

86. Current densities.—The current density in early alternator armatures was often very high; not more than 300 circular mils per ampere being allowed in many cases. Such armatures usually ran very hot at full load. The current densities used in modern machines are much lower, from 500 to 700 c. mils per ampere being allowed, as in the case of direct current machines. The armature conductor is usually of ordinary cotton covered magnet wire in the smaller machines, and when a conductor of considerable cross section is required a number of wires are grouped in multiple. In larger machines copper bars are frequently used, as these admit of a large cross section being put in a minimum space. Wire of rectangular cross section and copper ribbon are also used in some cases.

88. Outline of alternator design.—An alternator is usually designed to give an e. m. f. of prescribed value and frequency, and to be capable of delivering a prescribed current without undue heating. To design an alternator is to so proportion the parts as to satisfy the following conditions.

(*a*) The product of revolutions per second into the number of pairs of field magnet poles must equal the prescribed frequency according to equation (18).

(*b*) Equation (59) namely

$$E = \frac{4.44 \, kNTf}{10^8}$$

must be satisfied, to give the prescribed e. m. f.

(*c*) Peripheral speed of armature must not exceed allowable limits.

(*d*) The armature must have sufficient surface to radiate the total watts lost in the armature (including eddy surrent and hysteresis losses) without excessive rise of temperature. There are two distinct cases in the designing of an alternator as follows:

Case I.—Where the speed is fixed by independent considerations, as, for example, in direct connected machines. In this case the number of poles is determined by the given speed and prescribed frequency. The diameter of the armature follows from the allowable peripheral speed.* Assuming from 2% to 5%† of total rated output as armature loss, the approximate length of the armature is then determined by the radiating surface required. A well ventilated armature should radiate from .05 to .06 watt per square inch (of cylindrical surface) per degree Centigrade rise of temperature. This constant of radiation varies greatly with the style of construction of the armature and with peripheral speed. The length may be slightly modified when condition (*b*) comes to be considered. The flux N is de-

---

* Direct connected dynamos are scarcely ever run up to the allowable peripheral speed. Speeds from 2,200 to 2,600 feet per minute are usual.

† According to size of machine, low percentage being for large machines.

termined from the flux density in the air gap and the area of each pole-face. The combined area of the pole-faces is usually about equal to half the cylindrical surface of the armature, or, in other words, the distance between tips of adjacent poles is equal to the breadth of the pole face. The number of armature turns T is then determined from equation (59). The armature turns thus determined may come out an odd or a fractional number and must be adjusted to suit the type of winding employed, that is, to give the required number of coils each having the same number of turns. The length of the armature may then be changed slightly to adjust N so that the required e. m. f. will be produced with the adopted number of armature turns. The area of cross-section of the armature conductors is fixed by the allowable current density and rated current output.

Case II.—When speed is not fixed by independent considerations. In this case a trial combination of poles and speed is adopted giving a speed suitable for the size of the armature. The remainder of the design is then worked out as above.

Remark: When a machine is provisionally designed the details of its behavior may be approximately calculated without difficulty, and refinement of design is attained by working out a number of provisional designs and calculating the details of their action, then the most satisfactory design may be recognized and adopted.

The proportioning of the magnetic circuits and the calculation of field windings of an alternator is carried out in the same general way as in the case of a direct-current dynamo.

Remark: In designing a two-phase alternator each winding is allowed to cover half of the armature surface. In designing a three-phase alternator each winding is allowed to cover one-third of the armature surface.

89. Field excitation of alternators.—The use of an auxiliary direct-current dynamo for exciting the field of an alternator has been pointed out in Art. 19. The e. m. f. of an alternator excited in this way, falls off greatly with increasing current output and to counteract this tendency an auxiliary field excitation is frequently provided which increases with the current output of the machine. For this purpose the whole or a portion of the current given out by the machine is rectified * and sent through the auxiliary field coils.

Fig. 106 shows an alternator A with its field coils F separately

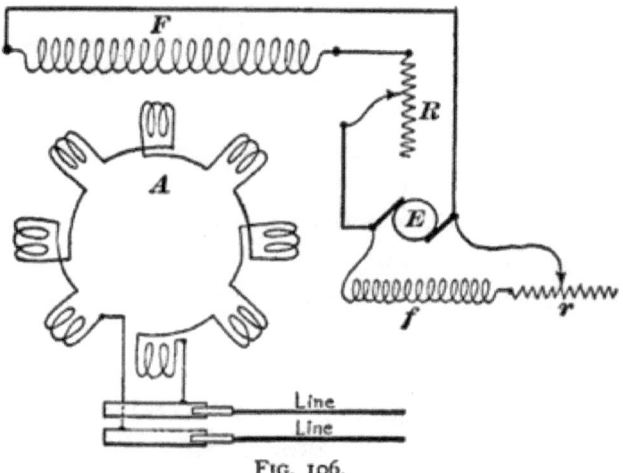

FIG. 106.

excited from a direct-current dynamo E. The two rheostats R and r, in series with the alternator and exciter fields respectively, are used to regulate the field current. Fig. 107 shows an alternator with two sets of field coils F and C. The coils F are separately excited as before. The coils C, known as the series or compound coils, are excited by current from the armature of the alternator. One terminal of the armature winding is connected directly to a collecting ring. The other armature terminal connects to one set of bars in the rectifying commutator B. From the rectifier the current is led through the winding C, thence back to the rectifier and thence to the second collecting ring. The rectifying commutator B is provided with as many segments

---

* Connections to field coils are reversed with every reversal of main current so that, in the field coils, the current is unidirectional.

as there are poles on the machine. This commutator reverses the connections of the terminals of the coils C at every pulsation of the alternating current so that the current flows in C always in the same direction. The commutator B is fixed to the armature shaft. A shunt s moving with the commutator is sometimes used when it is desired to rectify only a portion of the current. A stationary shunt s' is also frequently used to regulate the amount of current flowing around the coils C, thus giving a method of adjusting the compounding.

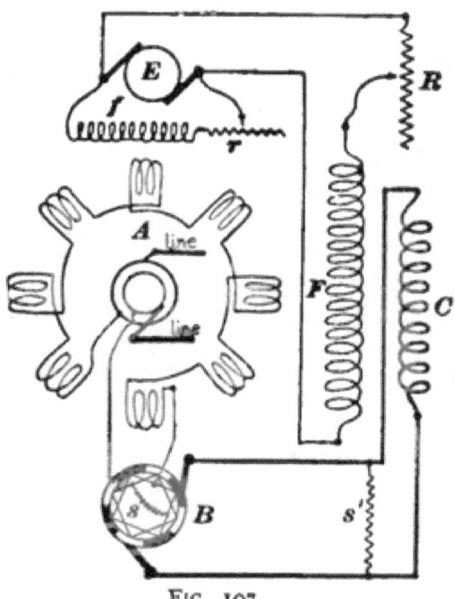

FIG. 107.

Fig. 108 shows an alternator A with two sets of field coils F and C as before. One armature terminal is connected to a collecting ring and the other armature terminal connects to the primary of a transformer T and thence to the other collecting ring. The terminals of the secondary coil of T connect to the bars of the rectifying commutator B from which the compound field winding C is supplied. The transformer T is usually placed inside the armature. All three of the methods shown in Figs. 106,

107 and 108 are in common use for the field excitation of alternators. Compounding is necessary only with alternators which

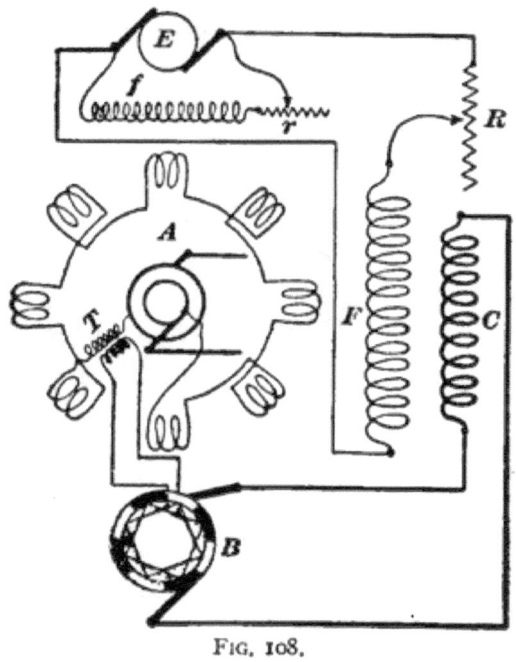

Fig. 108.

have fairly high armature inductance and which, with constant field excitation, would give poor regulation. For low inductance machines the separate excitation alone is usually sufficient.

CHAPTER X.

THE TRANSFORMER.

90. The transformer consists of an iron core upon which two separate and distinct coils of wire are wound. When one of these coils receives alternating current from any source the other coil delivers alternating current to any circuit which may be connected to its terminals. The coil which receives current is called the *primary* and the coil which delivers current is called the *secondary*. The alternating current in the primary coil magnetizes and demagnetizes the iron core repeatedly thus inducing an alternating e. m. f. in the secondary coil.

For the purpose of the immediately following discussion the resistances R' and R'' of the respective coils are supposed to be negligibly small; the magnetic reluctance of the core is supposed to be negligibly small so that a very weak magnetomotive force may produce a very large flux through the core; and all the magnetic flux N which passes through the primary coil is supposed to pass through the secondary coil also. A transformer satisfying these conditions would be called an *ideal transformer*.

91. Ratio of transformation.—Let an alternating e. m. f. (not necessarily harmonic) of which the instantaneous value is e' act on the primary coil of a transformer. Since R' is negligible this e. m. f. is all balanced by the counter e. m. f., $Z'\dfrac{dN}{dt}$, which is induced in the primary coil by the changing core flux N. Therefore

$$e' = Z'\frac{dN}{dt} \qquad (60)$$

in which Z' is the number of turns of wire in the primary coil. The changing flux induces in the secondary coil an alternating e. m. f. of which the instantaneous value is:

$$e'' = -Z'' \frac{dN}{dt} \qquad (61)$$

in which Z'' is the number of turns of wire in the secondary coil. This e. m. f. e'' is all available at the terminals of the secondary coil since R'' is negligible. Equations (60) and (61) show that e'' is at each instant equal to $\frac{Z''}{Z'} e'$ so that

$$\frac{E'}{E''} = \frac{Z'}{Z''} \qquad (62)$$

in which E' and E'' are the effective primary and secondary e. m. f.'s respectively.

When the circuit of the secondary coil is open the current in it is zero and the only current in the primary coil is the negligible current required to magnetize the core. When the secondary circuit is closed a current flows in it. Let i'' be the instantaneous value of this current then the instantaneous value i' of the primary current is such that

$$Z'i' + Z''i'' = 0. \qquad (63)$$

That is, the primary ampere turns, $Z'i'$, are equal and opposite to the secondary ampere turns $Z''i''$ (reluctance of core being negligible). Equation (63) shows that $\frac{i'}{i''}$ is always equal to $\frac{Z''}{Z'}$. Therefore

$$\frac{I'}{I''} = \frac{Z''}{Z'} \qquad (64)$$

in which I' and I'' are the effective values of primary and secondary currents respectively.

In an ideal transformer, then, the secondary e. m. f. E'' is opposite to the primary e. m. f. E' in phase and $\frac{Z''}{Z'}$ times as great; and the primary current I' is opposite to the secondary current I'' in phase and $\frac{Z''}{Z'}$ times as great. Therefore the angular phase difference θ between secondary e. m. f. and secondary current is equal to the angular phase difference between primary e. m. f. and primary current; and the power intake of the primary $E'I'\cos\theta$, is equal to the power output of the secondary $E''I''\cos\theta$.

The ratio $\dfrac{Z''}{Z'}$ is called the ratio of transformation of the transformer.

92. Particular cases (*for harmonic c. m. f. and current*). 1. *Non-inductive receiving circuit.*—In this case E'' and I'' are in phase and therefore E' and I' are in phase also. The state of affairs is represented in Fig. 109. The line ON represents the

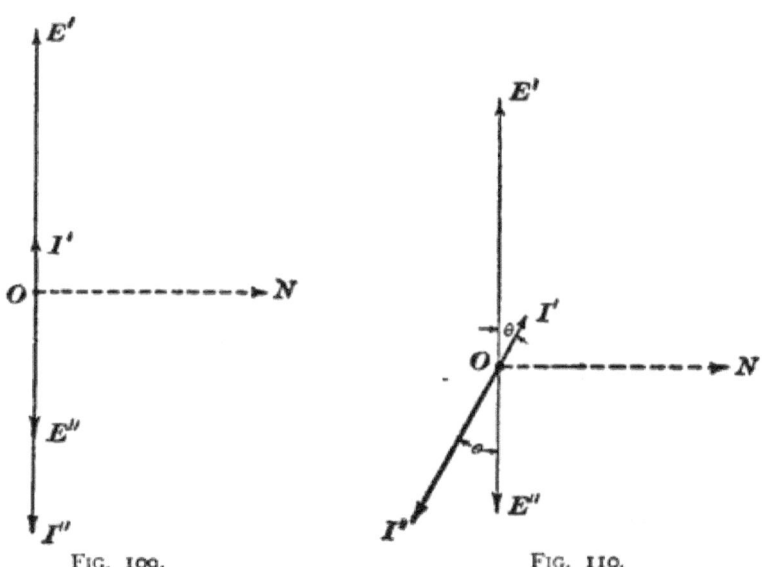

Fig. 109.   Fig. 110.

harmonically varying core flux, OE' represents the e. m. f. acting on the primary, and OE'' represents the e. m. f. induced in the secondary.

2. *Inductive secondary receiving circuit.*—In this case I'' lags behind E'' by the angle whose tangent is $\dfrac{\omega L}{R}$, where L^* is the in-

---

* The inductance of the secondary coil itself is already accounted for in the action of the transformer.

ductance and R the resistance of the receiving curcuit. Also I' lags behind E' by the same angle. The state of affairs is shown in Fig. 110.

3. *Secondary receiving circuit containing a condenser.*—In this case I'' is ahead of E'' by the angle whose tangent is $\left(\dfrac{1}{\omega J} - \omega L\right) \div R$, where J is the capacity of the condenser, L is the inductance of the connecting wires and R their resistance.

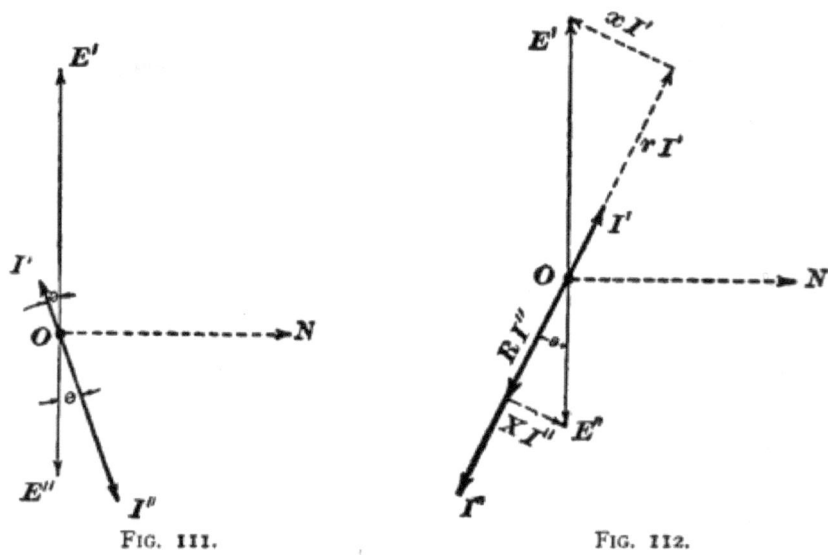

FIG. 111.   FIG. 112.

Also, I' is ahead of E' by the same angle. The state of affairs is shown in Fig. 111.

93. Equivalent resistance and reactance of an ideal transformer feeding a given secondary circuit.—The primary of a transformer takes from the mains a definite current at a definite phase lag when the secondary is supplying current to a given circuit. Consider a simple circuit of resistance r and reactance x which, connected to the mains, takes the same current as the primary of the transformer and at the same phase lag. This circuit is said to be equivalent to the transformer and its secondary receiving circuit, and r and x are called the equivalent primary resistance and reactance respectively of the secondary receiving circuit.

Resolve the primary e. m. f. E', Fig. 112, into components parallel to I' and perpendicular to I' as shown. The component parallel to I' is rI' and the component perpendicular to I' is xI' according to problem IV. The triangle whose sides are E', rI' and xI' is similar to the triangle whose sides are E'', RI'' and XI'', R and X being the given resistance and reactance of the secondary receiving circuit. Therefore

$$\frac{xI'}{XI''} = \frac{E'}{E''}$$

and

$$\frac{rI'}{RI''} = \frac{E'}{E''}$$

But $\dfrac{E'}{E''} = \dfrac{Z'}{Z''}$ and $\dfrac{I'}{I''} = \dfrac{Z''}{Z'}$ so that:

$$r = \left(\frac{Z'}{Z''}\right)^2 R \tag{65}$$

$$x = \left(\frac{Z'}{Z''}\right)^2 X. \tag{66}$$

That is, a transformer supplying current from its secondary to a circuit of resistance R and reactance X, takes from the mains the same current at the same phase lag as would be taken by a circuit of resistance $\left(\dfrac{Z'}{Z''}\right)^2 R$ and of reactance $\left(\dfrac{Z'}{Z''}\right)^2 X$ connected directly to the mains.

94. Maximum core flux.—When the e. m. f. acting on the primary of a transformer is harmonic there is a simple and important relation between E', ω, Z' and the maximum core flux N. Let e' be the instantaneous value of the primary e. m. f. Then since e' is harmonic we have

$$e' = E' \sin \omega t \tag{67}$$

in which E' is the maximum value of e'. Substituting the value of e' from (67) in (60) we have

$$\frac{dN}{dt} = \frac{E'}{Z'} \sin \omega t$$

whence by integration:

$$N = -\frac{E'}{\omega Z'} \cos \omega t. \tag{68}$$

That is, the core flux N is a harmonically varying quantity and its maximum value is:

$$N = \frac{E'}{\omega Z'}$$

or
$$N = \frac{\sqrt{2}E'}{\omega Z'}. \tag{69}$$

95. The actual transformer.—The action of the actual transformer deviates from the above described ideal action on account of the resistances R' and R'' of the coils; on account of eddy currents and hysteresis; and on account of the fact that some lines of flux pass through one coil without passing through the other (magnetic leakage).

However, a well designed transformer at moderate load approximates so closely to the ideal transformer in its action that equations (62) to (69) may be used in practical calculations.

In the following discussion the effects of core reluctance and eddy currents, of coil resistances, and of magnetic leakage are considered separately. These effects are small in themselves and their influence on each other is entirely negligible.

96. Effects of core reluctance and eddy currents upon the action of a transformer.—The magnetizing action of the primary current on the core of a transformer exceeds the opposite magnetizing action of the secondary current by an amount sufficient to balance the demagnetizing action of the eddy currents in the core and to overcome the magnetic reluctance of the core. Therefore, the primary current may be considered in three parts, namely: the part a which counteracts the magnetizing action of the secondary current; the part m the magnetizing action of which overcomes the reluctance of the core; and the part c which counteracts the magnetizing action of the eddy currents. Thus OA, Fig. 113, represents the part a of the primary current, this part is opposite to I'' in phase and equal to $\frac{Z''}{Z'} \times I''$; OM represents the part m; and OC represents the part c. The line OI' which is the vector sum of OA, OM and OC represents the total primary current. The currents m and c are usually small compared with the full load primary current of a transformer. The currents m and c together, namely $m + c$, constitute what is called the *leakage* or *magnetizing* current of the transformer.

Discussion of the part c of the primary current.—This part of the primary current is harmonic when the primary e. m. f. is harmonic and it is in phase with the primary e. m. f.* Let C be the

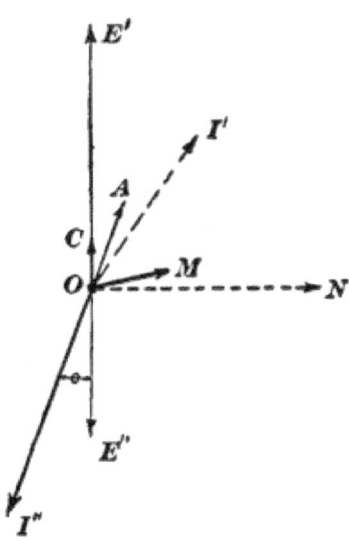

FIG. 113.

effective value of the current c; then the power taken from the mains by the current c; is $E'\,C$ and this power is equal to the eddy current loss, W_e, in the core, Therefore

$$C = \frac{W_e}{E'}. \qquad (70)$$

The method of calculating eddy current loss W_e is given later.
Discussion of the part m of the primary current.—This part of

---

* *Proof:* The primary e. m. f. and the eddy currents are at their maximum value at the instant when the core flux is changing fastest. That is eddy currents are in phase with primary e. m. f. (really in opposition to primary e. m. f.). The current c which opposes the magnetizing action of the eddy currents is of course opposite to eddy currents in phase and therefore in phase with primary e. m. f.

the primary current is not harmonic, but since it is small it may be treated as a harmonic current without great error. Let M be the effective value of the current m; let M' be the component of M parallel to E'; and let M'' be the component of M perpendicular to E'. Then $E'M'$ is the power taken from the mains by the current m and this power is equal to the hysteresis loss, W_h, in the core. Therefore

$$M' = \frac{W_h}{E'}. \qquad (a)$$

The component M'' of the current m reaches its maximum value $\sqrt{2}M''$ when the core flux is at its maximum value N and inasmuch as this component of m overcomes the reluctance of the core we have

$$N = \frac{\frac{4\pi}{10} Z' \sqrt{2} M''}{G}$$

or

$$M'' = \frac{10NG}{4\sqrt{2}\pi Z'} \qquad (b)$$

in which G is the magnetic reluctance of the core.

Total leakage current.—From equations (70), (a) and (b) we have

$$\text{Power component of leakage current} = \frac{W_h + W_e}{E'} \qquad (71)$$

$$\text{Wattless component of leakage current} = \frac{10NG}{4\sqrt{2}\pi Z'} \qquad (72)$$

in which G is the magnetic reluctance of the core, Z' is turns of wire in primary coil, and N is the maximum value of the core flux.

97. Actual value of the part m of the primary current.

(*a*) *When the core is assumed to be without hysteresis.*—Let the ordinates of the curve mN, Fig. 114, represent values of core flux N produced by various given current strengths m in the primary coil, these current strengths being represented by the abscissas of the curve mN.

When the primary e. m. f. is harmonic then the core flux N is harmonic also, and 90° behind E' in phase, according to equation (68). Let the sine curve N, Fig. 114, represent the value of N as time passes; time as abscissas and N as ordinates.

Then the curve m of which the ordinates represent successive instantaneous values of the current m is constructed as follows: Draw the ordinate dp and the abscissa ap. Lay of dc equal to ab which is the magnetizing current required to force through the core the flux dp. The locus of the point c is the required curve. The figure shows that the magnetizing current is not harmonic although it is wattless.

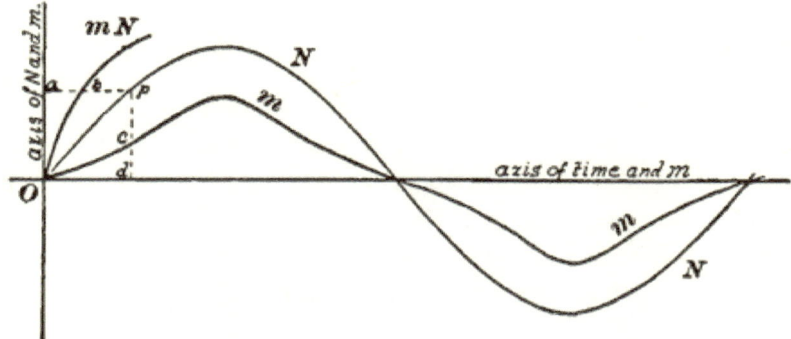

FIG. 114.

(b) *When the hysteresis is taken into account.*—Let the ordinates of the curve mN, Fig. 115, represent values of core flux produced by various given current strengths in the primary coil, these current strengths being represented by the abscissas of the curve mN. The curve m of which the ordinates represent the successive instantaneous

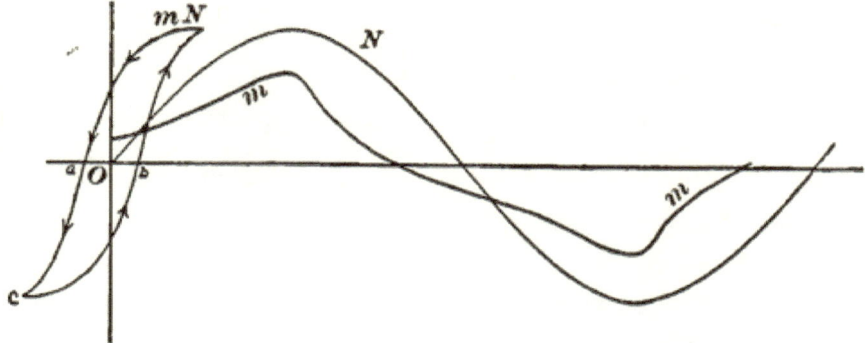

FIG. 115.

values of the current m is constructed as before; the ascending branch of the hysteresis loup mN being used for increasing values of N and the descending branch for decreasing values of N.

98. Transformer regulation. *Preliminary statement concerning the effects of magnetic leakage and of resistances of primary* *and secondary coils on the action of a transformer.*—In the ideal transformer the whole of the primary e. m. f. is balanced by the opposite e. m. f. induced in the primary coil by the varying magnetic flux which passes through both coils, and the whole of the e. m. f. induced in the secondary coil by this varying flux is available at the terminals of the secondary coil.

In the actual transformer a portion of the primary e. m. f. is lost in overcoming the resistance of the primary coil and a portion is lost in balancing the e. m. f. which is induced in the primary coil by the flux which passes through the primary coil but does not pass through the secondary coil (leakage flux). These lost portions of the primary e. m. f. are proportional to the primary current so that the useful part * of the primary e. m. f. falls short † of the total primary e. m. f. by an amount which is proportional to the current.

The total e. m. f. induced in the secondary coil is proportional to the *useful part* of the primary e. m. f. and a portion of the total secondary e. m. f. is lost in overcoming the resistance of the secondary coil. This lost portion of the secondary e. m. f. is proportional to the secondary current (or to primary current, since the ratio of the currents is constant).

Therefore the effect of magnetic leakage and of coil resistances is to make the e. m. f. between the terminals of the secondary coil fall short † of its ideal value $\dfrac{Z''}{Z'} \cdot E'$ by an amount which is proportional to the current.

This falling off of secondary e. m. f. with increasing current is of practical importance inasmuch as most receiving apparatus must be supplied with current at approximately constant e. m. f. A transformer of which the secondary e. m. f. falls off but little

---

* The part, namely, which balances the e. m. f. induced in the primary coil by the magnetic flux which passes through both coils.

† The lost portions of primary and secondary e. m. f.'s are in general not in phase with total primary and total secondary e. m. f.'s. These losses are therefore to be subtracted as vectors as explained in the following articles.

with increase of current is said to have good regulation. A transformer to regulate well must have low resistance coils and little magnetic leakage. Large transformers as a rule regulate more closely than small ones.

99. Effect of resistance of coils upon the action of a transformer.—Fig. 116 shows the general effect of the resistances of the coils upon the action of a transformer. The line ON represents the harmonically varying flux in the core. Oa represents the useful part of the primary e. m. f. and Ob the total e. m. f. induced in the secondary coil. The line OI'' represents the secondary current and the line OI' represents the primary current. The total primary e. m. f. E' exceeds Oa by the amount $R'I'$ (parallel to I'), and the e. m. f. E'' at the terminals of the secondary coil falls short of Ob by the amount $R''I''$ (parallel to I'').

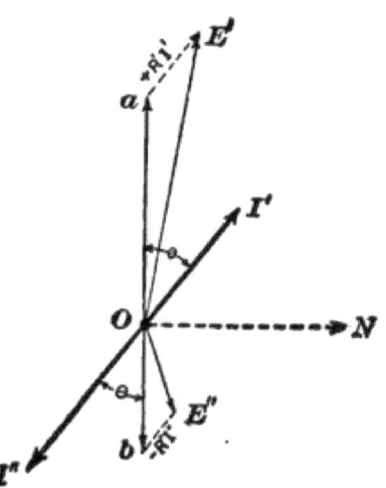

FIG 116.

Remark: When the angle θ, Fig. 116, is nearly zero (secondary receiving circuit noninductive) then $R'I'$ and $R''I''$ are nearly parallel to Oa and Ob respectively, so that Oa is much less than E' in value and E'' is much less than Ob in value. On the other hand, when the angle θ is nearly $\pm$ 90° (secondary receiving circuit containing a large inductance or a condenser) then $R'I'$ and $R''I''$ are nearly perpendicular to Oa and Ob respectively, so that Oa is nearly equal to E' in value and E'' is nearly equal to Ob in value. Therefore the regulation of a transformer is largely affected by coil resistance when the secondary receiving circuit is noninductive, but scarcely at all affected by the coil resistance when the secondary receiving circuit contains a large inductance or a condenser.

100. Effect of magnetic leakage upon the action of a transformer.—It is shown in the next article that magnetic leakage is in its effects equivalent to an auxiliary outside inductance P through which the primary current passes on its way to the primary of the transformer. The part of the primary e. m. f. L' which is lost in this inductance is equal to $\omega PI'$ and it is 90° ahead of the primary current I' in phase.

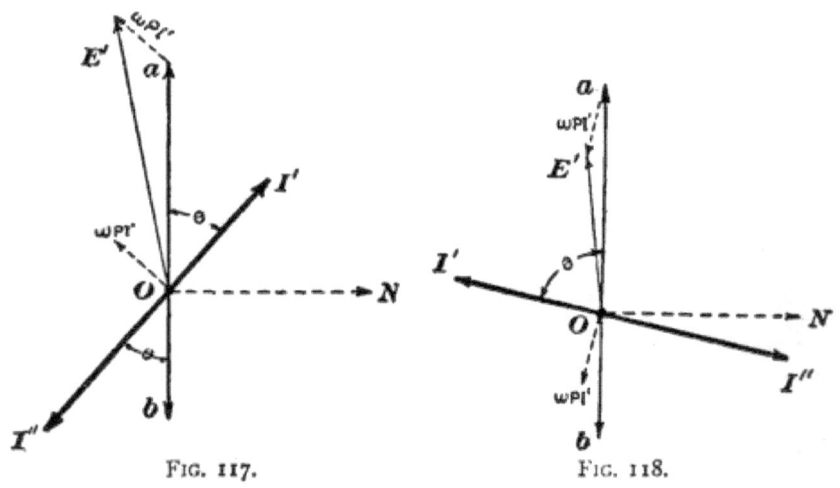

FIG. 117.   FIG. 118.

When the secondary receiving circuit is inductive I'' lags behind E'' (Ob, Fig. 117) by the angle θ as shown in Fig. 117, and the useful part, Oa, of the primary e. m. f. is *less* than the total primary e. m. f. In this case the secondary e. m. f., which is equal to $\dfrac{Z''}{Z'} \times Oa$, falls off in value as I' (and also $\omega PI'$) increases.

When the secondary receiving circuit contains a condenser I'' is ahead of E'' (Ob, Fig. 118) as shown in Fig. 118, and the useful part, Oa, of the primary e. m. f. is *greater* than the total primary e. m. f. in value. In this case the secondary e. m. f., which is equal to $\dfrac{Z''}{Z'} \times Oa$, increases in value as I' (and also $\omega PI'$) increases.

THE TRANSFORMER. 131

When the secondary receiving circuit is noninductive the angle θ is zero and $\omega PI'$ is at right angles to Oa so that Oa is sensibly equal to E' in value and therefore sensibly constant. In this case the secondary e. m. f. remains sensibly constant as I' (and also $\omega PI'$) increases.

101. Proposition. *The effect of magnetic leakage in a transformer is equivalent to a certain outside inductance P, connected in series with the primary coil.*

Discussion.—Let A, Fig. 119, be the primary coil, B the secondary coil, and C the iron core of a transformer. As the (harmonic) alternating currents in A and B pulsate, harmonically varying fluxes are produced through the core and around the coils. Let OC, Fig. 120, represent the harmonically varying flux through the core; Op the harmonically varying flux which encircles coil A only; and Os the harmonically varying flux which encircles coil B only.

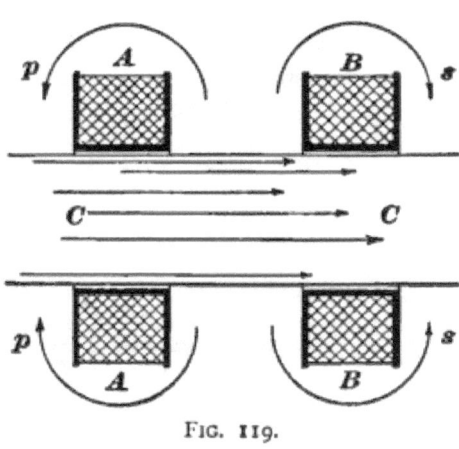

FIG. 119.

The fluxes Op and Os are proportional to and in phase with I' and I'' respectively, so that the total flux $Op + Os$ (represented by the lines sp or ba, Fig. 120) which passes between A and B is proportional to and in phase with I'.

The total harmonically varying flux through coil A is $OC + Op$ or Oa, and the total harmonically varying flux through coil B is $OC + Os$ or Ob. Now, $Oa = Ob + ba$, so that we may look

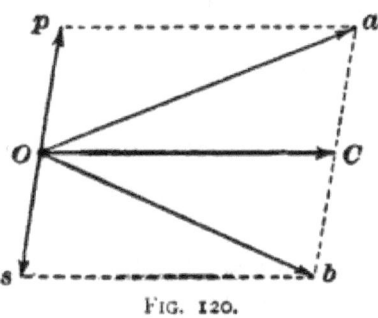

FIG. 120.

upon the action of the transformer as due to the flux Ob passing through both coils and the flux ba passing through the primary coil only. This latter flux being proportional to the primary current is equivalent in its effects to an inductance P, connected in series with the primary coil. Let N' be the value of the leakage flux ab which, for a given value i' of the primary current, encircles the primary coil, then, according to equation (5) we have

$$P = \frac{Z'N'}{i'}. \qquad (73)$$

102. Calculation of leakage inductance P.—The leakage flux N' equation (73) [$= ba$, Fig. 120], and therefore the value of P also, depends upon the size and shape of the primary and secondary coils and upon their proximity to the core. In considering the flux between the coils (leakage flux) we need not consider whether a given portion of this flux is a part of Op or a part of Os, Fig. 120, inasmuch as these two fluxes are added together to give ba or N'.

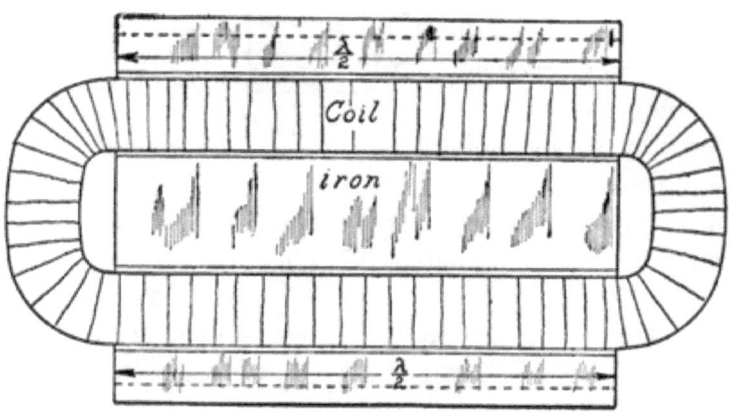

Fig. 121.

Figs. 121 and 122 show side and end views of a shell type transformer. The trend of the leakage flux is shown in the upper part of Fig. 122 (omitted from lower part for the sake of clearness), and the dimensions X, Y, g, λ and l are shown. Fig.

123 is an enlargement of the upper part of Fig. 122. Consider the flux across between the dotted lines aa, Fig. 123. The magnetomotive force pushing this flux across is $4\pi i' \frac{x}{X} Z'$.* The length of the air portion of the magnetic circuit through which the leakage flux flows is l and its sectional area is λdx (counting both limbs of the coils). Therefore, the magnetic reluctance of this leakage circuit is $\frac{l}{\lambda \cdot dx}$ and the flux across between aa is $\frac{m.\,m.\,f.}{m.\,r.} = 4\pi i' Z' \frac{\lambda x dx}{lX}$.

This flux encircles the fractional part $\frac{x}{X}$ of the primary turns and, therefore, the fractional part $\frac{x}{X}$ of the flux is to be counted as encircling the entire primary coil so that

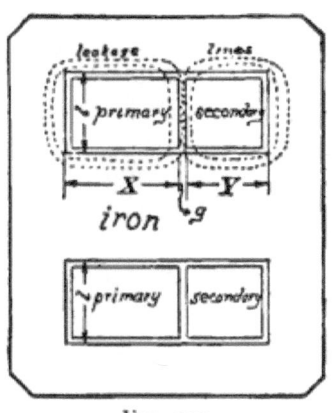

FIG. 122.

$$dN' = \frac{4\pi Z' i' \lambda}{lX^2} \cdot x^2 dx.$$

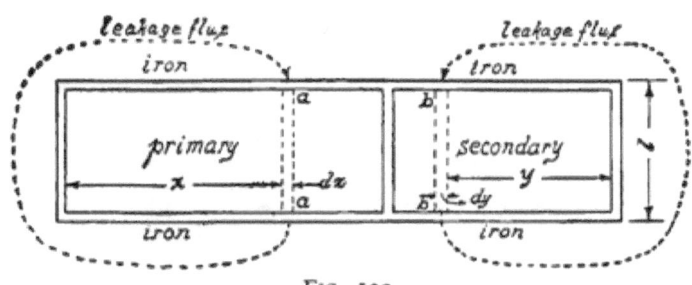

FIG. 123.

The part of N' which flows across the primary coil is the integral of this expression from $x = 0$ to $x = X$. This part of N' is therefore

---

* All quantities in this article are expressed in c. g. s. units.

$$\frac{4\pi Z'i'\lambda X}{3l}.$$

Similarly, the part of N' which flows across the secondary coil is

$$\frac{4\pi Z'i'\lambda Y}{3l}.^*$$

The flux across the gap g between the primary and secondary coils is all counted as a part of N' and is equal to

$$\frac{4\pi Z'i'\lambda g}{l}.$$

Therefore

$$N' = \frac{4\pi Z'i'\lambda}{l}\left[\frac{X}{3} + \frac{Y}{3} + g\right]. \qquad (74)$$

There is some leakage flux passing between the primary and secondary coils where they project beyond the iron core. This part of the leakage flux has a longer air path than the leakage flux which flows from iron to iron, say three times as long. Therefore, for λ we may take the total length of the coils lessened by say ⅔ the length which is surrounded by air only.

Substituting the value of N' from (74) in (73) we have

$$P = \frac{4\pi Z'^2\lambda}{l}\left[\frac{X}{3} + \frac{Y}{3} + g\right]. \qquad (75)$$

This equation gives the value of P in centimeters, all dimensions being expressed in centimeters.

The equivalent inductance P may be reduced in value by lessening λ, X, Y or g or by increasing l. Fig. 124 shows the proportions of a recent type of transformer for which the leakage inductance P is very small. The value of P may be further reduced by winding the primary and secondary coils in alternate sections.

103. Calculation of transformer regulation.—Let P be the inductance equivalent of magnetic leakage of a transformer and let

---

* In this expression $Z'i'$ being equal to $Z''i''$ is written therefor.

$$R''' = R' + \left(\frac{Z'}{Z''}\right)^2 R''$$

where R''' is the primary resistance which is equivalent in its effects to the resistance R' of the primary coil and the resistance R'' of the secondary coil combined. Let x and r be the primary equivalents of the reactance X and resistance R respectively of the secondary receiving circuit. [See equations (65) and (66).] Then, aside from the negligible effects of hysteresis and eddy currents on regulation, the transformer and its receiving circuit are equivalent

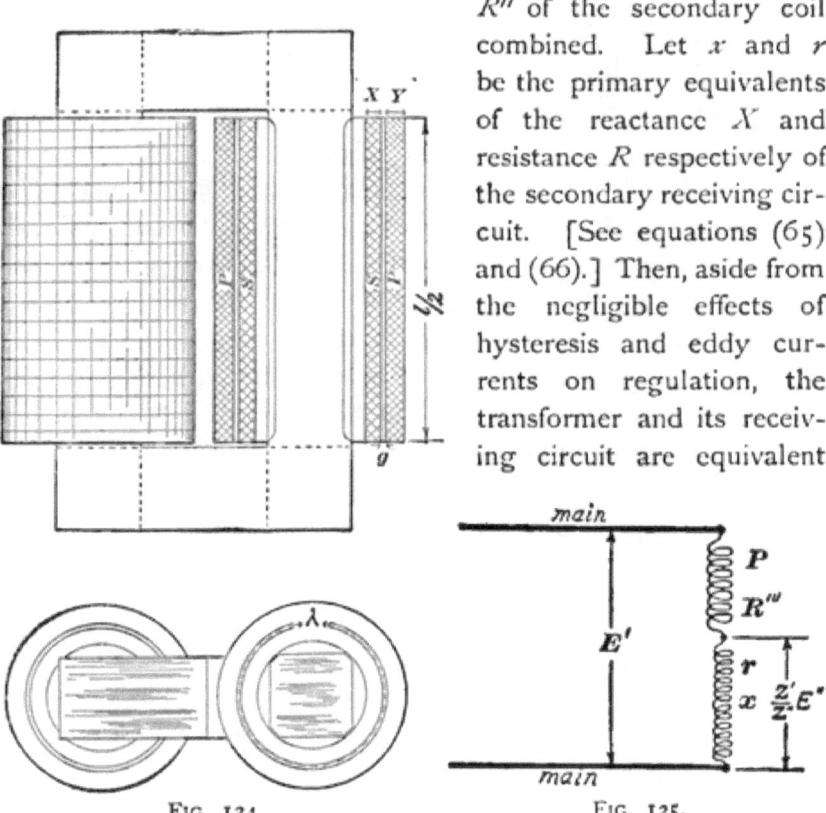

FIG. 124.   FIG. 125.

to the circuit shown in Fig. 125. The problem of transformer regulation is thus reduced to the problem of two coils in series. See Problem VII., Chapter VII.

104. The constant current transformer.—A transformer of which the leakage inductance P is very large is sometimes called a *constant current* transformer for the reason that the current delivered by such a transformer varies but little with the resistance of the receiving circuit, so long as this resistance is comparatively small, the primary of the transformer being connected to constant e. m. f. mains. The action of the inductance P in controlling the current is explained in the article on the constant current alternator. (Art. 78.)

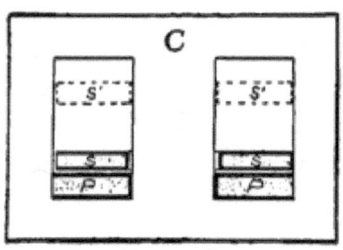

FIG. 126.

Fig. 126 is a sketch of the General Electric Company's type of constant current transformer; C is the iron core, PP the primary coil and SS the secondary coil. The secondary coil is movable and nearly counterbalanced, and the increased repulsion between PP and SS due to a slight increase of current lifts the secondary coil to $S'S'$.

When the primary and secondary coils are near together the leakage inductance is very small and a decrease in the resistance of the receiving circuit would be accompanied by a great increase of current were it not for the movement of the secondary coil and the consequent increase of leakage inductance.

CHAPTER XI.

TRANSFORMERS.
(*Continued.*)

105. Transformer losses.—The power output of a transformer is less than its power intake because of the losses in the transformer. These losses are: (*a*) The iron or core losses due to eddy currents and hysteresis; and (*b*) The copper losses due to the resistances of the primary and secondary coils.

The iron losses are practically the same in amount at all loads, and they depend upon the frequency and range of the flux density B, upon the quality and volume of the iron, and upon the thickness of the laminations.

The hysteresis loss in watts is

$$W_h = aVfB^{1.6}. \qquad (76)$$

Where f is the frequency in cycles per second, B is the maximum flux density in lines per square centimeter, V is the volume of the iron in cubic centimeters, and a is a constant depending upon the quality of the iron. For annealed refined wrought iron the value of a is about 3×10^{-10}.

The eddy current loss in watts is:

$$W_e = bVf^2l^2B^2. \qquad (77)$$

Where l is the thickness of the laminations in centimeters, and b is a constant depending upon the quality* of the iron. For ordinary iron the value of b is about 1.6×10^{-11}. Insufficient insulation of laminations causes excessive eddy current loss.

Remark: Equations (76) and (77) may be used for calculating the hysteresis and eddy current losses in any mass of laminated iron subjected to periodic magnetization, such as alternator armatures and the rotor and stator iron in an induction motor.

---

* Upon the specific resistance of the iron.

The copper loss is:

$$W_e = R'I'^2 + R''I''^2. \tag{78}$$

This loss is nearly zero when the transformer is not loaded; it increases with the square of the current, and becomes excessive when the transformer is greatly overloaded.

106. Efficiency of transformers.—The ratio power output ÷ power intake is called the efficiency of a transformer. The accompanying table shows the full-load efficiencies of various sized transformers of a recent type.

TABLE OF TRANSFORMER EFFICIENCIES.

Output Kilo-watts.	Per cent. Efficiency Full Load.
1	94.8
2	95.75
3	96.2
4	96.45
5	96.65
6	96.73
7	96.8
8	96.85
9	96.9
10	96.95
15	97.2

The efficiency of a given transformer is very low when the output is small, it increases as the output increases, reaches a maximum, and falls off again when the output is very great. This falling off of efficiency when the output is great is due to the great increase of copper losses. Fig. 127 shows the efficiency of a transformer at various loads.

Calculation of efficiency.—The transformer output (noninductive receiving circuit) is $E''I''$. The internal loss is $W_h + W_e + W_c$ so that the intake is $E''I'' + W_h + W_e + W_c$ and the efficiency is:

$$\eta = \frac{E''I''}{E''I'' + W_h + W_e + W_c}. \tag{79}$$

All-day efficiency.—Usually a transformer is connected to the mains continuously, and current is taken from the secondary for a few hours, only, each day. In this case the iron loss is incessant and the copper loss is intermittent. The total work given to the transformer during the day may greatly exceed the total work given out by it, especially if the incessant iron losses are not reduced to as low a value as possible. The ratio total work given out by the transformer ÷ total work received by the transformer during the day is called the all day efficiency of the transformer.

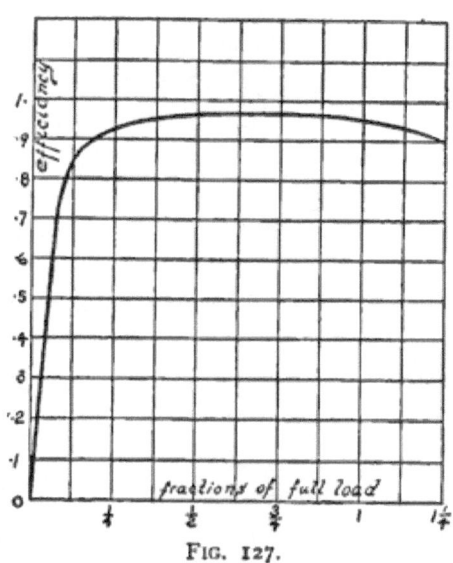

FIG. 127.

107. Practical and ultimate limits of output of a transformer.—When the secondary current of a transformer is increased the secondary e. m. f. generally drops off, and the power output increases with the current and reaches a maximum as in the case of the alternator. This maximum power output is the ultimate limit of output of the transformer. Practically the output of a transformer is limited to a much smaller value than this maximum output because of the necessity of cool running, because in most cases it is necessary that the secondary e. m. f. be nearly constant, and because the efficiency of a transformer is low at excessive outputs.

Small transformers have relatively large radiating surfaces and in such transformers the requirements of close regulation, as a rule, determine the allowable output.

Large transformers have relatively small radiating surfaces and their allowable output is limited by the permissible rise in temperature. Very large transformers are usually provided with air passages through which air is made to circulate by a fan. Sometimes transformers are submerged in oil, which, by convection, carries heat from the transformer to the containing case, where it is radiated.

Large transformers are much more efficient, under full load, than small ones, and give closer regulation.

108. Rating of transformers.—A transformer is rated according to the power it can deliver steadily to a noninductive receiving circuit without undue heating; and the ratio of transformation, together with a specification of the frequency and effective value of the primary e. m. f. to which the tranformer is adapted, are given.

The rating of a transformer is by no means rigid. Thus, if a transformer is used to give more than its rated output it will become somewhat more heated by the internal losses and its regulation will not be so close. If a transformer is used for a primary e. m. f. greater than its rated primary e. m. f. or for a frequency lower than its rated frequency, the range of flux density B in the core will be increased, which will increase the core losses. Some manufacturers rate their transformers generously, so that they may be greatly overloaded or used with greatly increased primary e. m. f. or decreased frequency without difficulty.

109. Outline of transformer design.—A transformer is usually designed to take current from mains at a prescribed e. m. f. and frequency, and to deliver current at a prescribed e. m. f. to a receiving circuit; the transformer must be so proportioned and of such size as to deliver the prescribed amount of current steadily without undue heating and without any great variation of its secondary e. m. f. from zero to full load.

In the designing of a transformer there is but one condition which must be precisely met, namely, the ratio of primary to secondary turns must be equal to the ratio of the prescribed primary and secondary e. m. f.'s. All other points in design are to a great extent matters of choice guided in a general way by experience.

The accompanying table gives magnetic flux densities which are usually employed in transformer cores.

The allowable temperature rise varies greatly with different makers, the extent of radiating surface required per watt of loss per degree rise of temperature varies between extremely wide limits, and no simple rule can be given covering this matter.

Magnetic Densities B for Transformer Cores.

Frequency.	Small Transformers.	Medium Size Transformers.	Large Transformers.
25	7500	6750	6000
40	6500	5750	5000
60	5000	4750	4500
80	4500	4250	4000
100	4000	3750	3500
120	3500	3250	3000

Given the required power output* of a transformer, the value and frequency of the primary e. m. f., and the value of the secondary e. m. f., the design of the transformer is conveniently determined as follows :

Find from the table the efficiency which can probably be attained and calculate the total transformer loss at full load. Of this total loss about half should be iron loss and half copper loss.† The total iron loss is :

$$W_i = aVfB^{1.6} + bVf^2l^2B^2 \qquad (80)$$

according to equations (76) and (77).

Having decided upon maximum flux density B (see accompanying table), and upon thickness‡ of laminations l, equation

---

* Rated output is the output which the transformer can deliver satisfactorily to a *noninductive* circuit.

† If the transformer is to be connected to the mains all day but is to deliver current only four hours per day, for example, then the iron loss during 24 hours should be about equal to the copper loss during four hours or under full load the copper loss should be several times as great as the iron loss.

‡ 12 to 16 thousandths of an inch is the thickness usually employed.

(80) gives the volume V of iron to be used in the transformer core. The core may be made of the type shown in Fig. 128 or of the type shown in Fig. 129. The proportions (relative dimensions) indicated in Figs. 128 and 129 will be found to give satis-

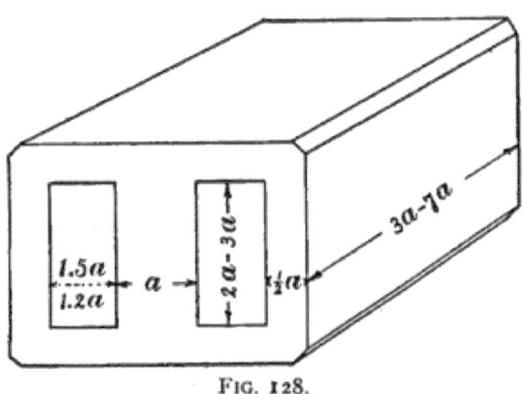

FIG. 128.

factory results although the form of the core may be considerably modified without greatly affecting the action of the transformer. In fact, it is usually necessary to modify the core slightly after the coils have been designed.

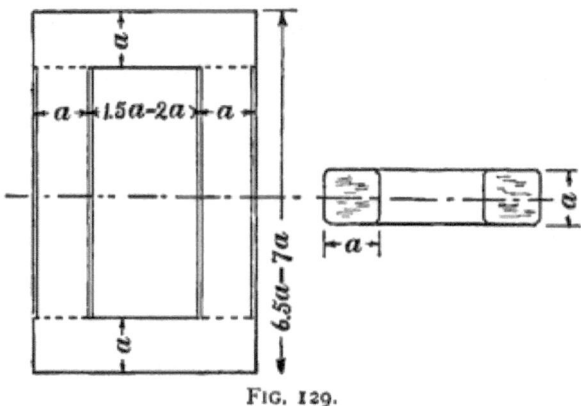

FIG. 129.

The maximum core flux N is equal to the product of the sectional area of the magnetic circuit (where it passes through the coils) into the maximum flux density B. Then equation (69) determines the primary turns, namely,

$$Z' = \frac{\sqrt{2}E' 10^8}{\omega N}$$

or
$$Z' = \frac{E' 10^8}{4.44 Nf} \qquad (81)$$

The number of secondary turns is then determined by equation (62) namely,

$$Z'' = \frac{E''}{E'} Z'. \qquad (82)$$

From the provisionally designed core the mean length of a turn of primary and of secondary coils may be determined which together with Z' and Z'' gives the total lengths of wire in primary and secondary. The size of this wire is then easily chosen so that $R'I'^2$ and $R''I''^2$ may be each equal to half the full load copper loss.

The size of wires being thus determined the space necessary for the coils and insulation can be estimated. If the provisionally designed core gives more or less space than is required for the coils its dimensions may be altered to suit.

Transformer Connections.

110. Simple connection. In parallel. In series.—When used to supply current to lamps or motors from constant potential

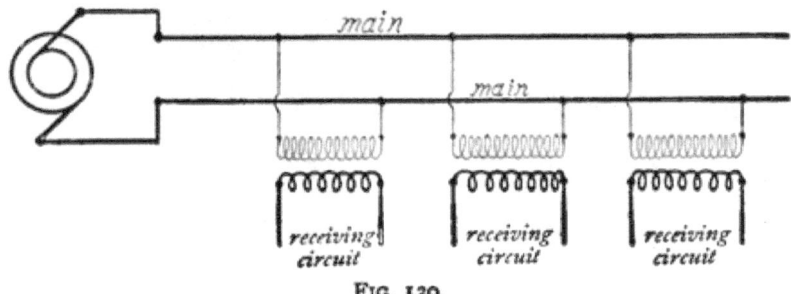

Fig. 130.

mains the primary of the transformer is connected to the mains and the secondary of the transformer is connected to the terminals of the receiving circuit. When a number of receiving circuits are supplied through separate transformers the primaries of the transformers are connected in parallel, as shown in Fig. 130.

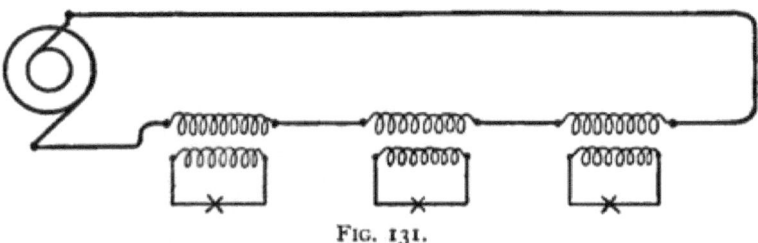

FIG. 131.

When current is supplied through transformers to a number of arc lamps from a constant current alternator the transformer primaries are connected in series and the lamps are connected to the respective secondaries, as shown in Fig. 131. This arrangement is seldom employed.

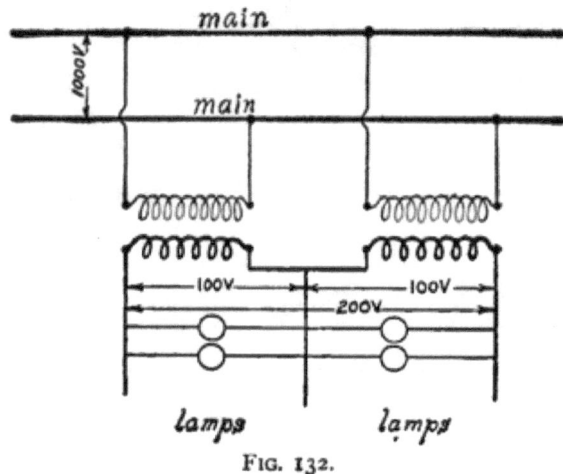

FIG. 132.

111. Transformers with divided coils.—Alternators for isolated lighting plants give usually 1000 or 2000 volts e. m. f. and the standard e. m. f.'s for incandescent lamps are 55 and 110 volts.

Transformers are frequently made with two primary coils, which may be connected in series for 2000 volts or in parallel for 1000 volts, and with two secondary coils, which may be connected in series to give 110 volts or in parallel to give 55 volts.

Transformers for supplying current for testing purposes are frequently made with a number of secondary coils, which may be connected to give high or low e. m. f.'s as desired.

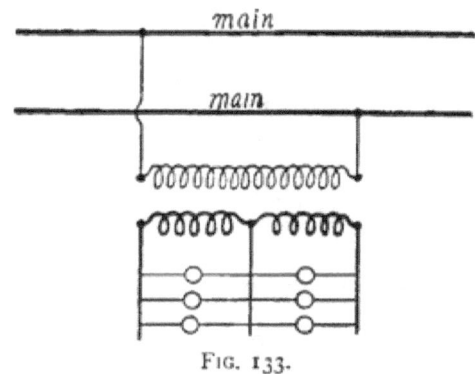

FIG. 133.

Transformers for supplying current to the Edison three-wire system.—For this purpose two similar transformers may be used as shown in Fig. 132, or a single transformer with two secondary coils may be used as shown in Fig. 133.

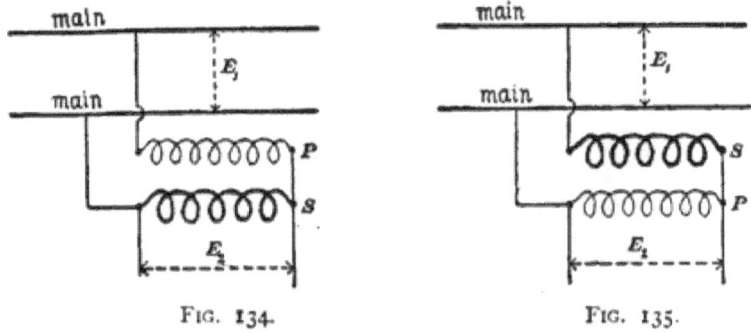

FIG. 134.   FIG. 135.

112. Connecting of primary and secondary coils in series.—The ratio of transformation of a given transformer may be altered by connecting the secondary coil in series with the primary as shown in Figs. 134 and 135. The arrangement shown in Fig. 134 gives a ratio of transformation $\frac{E_1}{E_2}$ equal to $a \pm 1$, and the arrangement shown in Fig. 135 gives $\frac{E_1}{E_2} = 1 \pm \frac{1}{a}$ where a is the ordinary ratio of transformation of the transformer. The + or — sign is taken according as the secondary is connected so as to help or so as to oppose the primary. These arrangements are not advisable in commercial work for the reason that they involve the connecting of the low e. m. f. mains directly to the high e. m. f. mains which is a source of danger.

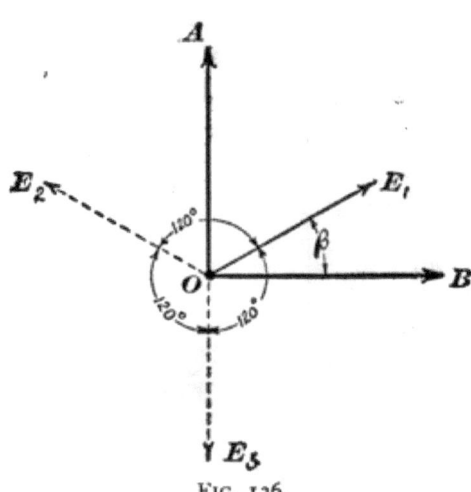

Fig. 136.

113. Two-phase three-phase transformers.—The use of simple transformers for two-phase and for three-phase systems has been explained. Transformers are frequently used to transform from two-phase to three-phase or *vice versa*.

To produce an e. m. f. of any specified value and phase.—Let A and B, Fig. 136, be the two e. m. f.'s of a two-phase dynamo and let it be required to produce an e. m. f. E_1 of given value and phase. The component of E_1 par-allel to A is $E_1 \sin \beta$ and the component of E_1 parallel to B is

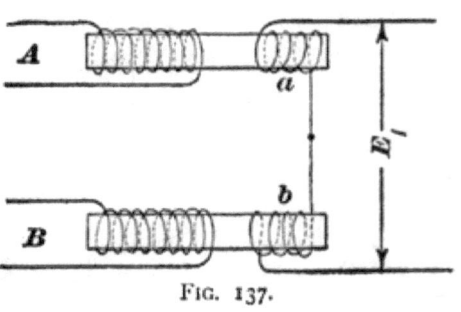

Fig. 137.

$E_1 \cos \beta$. Fig. 137 shows two distinct transformers with similar primary coils one connected to phase A the other to phase B. A secondary a may be wound upon the one transformer to give the component $E_1 \sin \beta$; and a secondary b may be wound upon the other transformer to give the other component $E_1 \cos \beta$. Similarly any other e. m. f. such as E_2 or E_3, Fig. 136, may be produced by a pair of properly proportioned secondary coils.

The two-phase three-phase transformer consists of two distinct transformers A and B, Fig. 137, wound with similar primary coils to which the two-phase e. m. f.'s are connected. Each of the three-phase e. m. f.'s is (in general) generated in a pair of secondary coils, one on each transformer. Such a transformer transforms equally well from three-phase to two-phase or

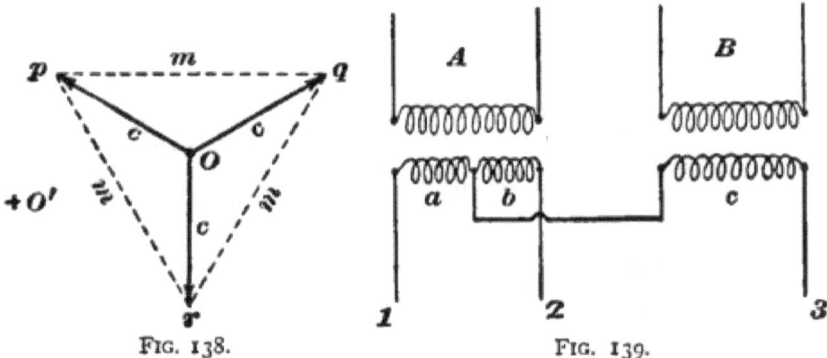

FIG. 138.   FIG. 139.

from two-phase to three-phase. The three pairs of secondary coils may be connected according to the Δ scheme or Y scheme. In the first case the e. m. f.'s between the three-phase mains are the e. m. f.'s produced in the respective pairs of coils. In the second case the e. m. f.'s between the mains (m) are related to the e. m. f.'s generated by the respective pairs of coils (c) as shown in Fig. 138 and as explained in Art. 68.

Consider any point O', Fig. 138. If a pair of secondary coils is arranged on A and B, Fig. 137, to give an e. m. f. $O'p$, another pair to give an e. m. f. $O'q$, and a third pair to give an e. m. f.

$O'r$; then these three pairs of coils Y-connected would give the three phase e. m. f.'s m m m between the mains.

Scott's transformer.—The simplest two-phase three-phase transformer is that due to Scott. This transformer consists of two cores with similar primaries A and B, Fig. 139. These two primaries are connected to the two-phase mains. One core has a single secondary coil c and the other has two similar secondaries, each having $\frac{1}{2}\sqrt{3}$ times as many turns as the coil c. These coils a b and c are Y-connected to the three-phase mains 1, 2 and 3, as shown. The point O', Fig. 138, lies for Scott's transformer midway between the points p and r as shown in Fig. 140. In this figure a b and c represent the e. m. f.'s induced in the coils a b and c, Fig. 139, respectively; the two-phase e. m. f.'s A and B being parallel to a and to c respectively as shown.

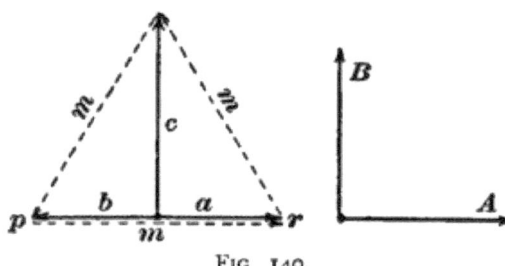

Fig. 140.

114. The monocyclic system.—The monocyclic generator of the Gen. Elec. Co. is a polyphase dynamo not strictly to be called two-phase or three-phase. It is employed in stations where a small portion of the output is used for motors and a large portion for lighting. The armature winding of the monocyclic generator is essentially a two-phase

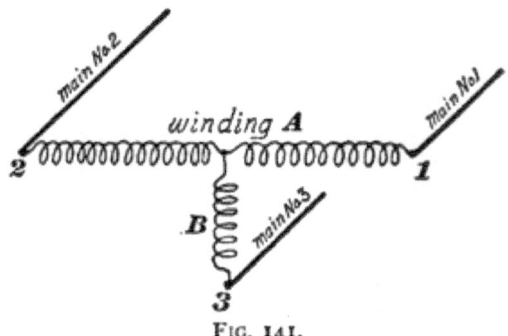

Fig. 141.

winding. The A winding has four times as many conductors as the B winding, and one end of the B winding is connected to the middle point of the *A* winding, as shown in Fig. 141. The three collecting rings are indicated by 1, 2 and 3. Main 3 is called the *teaser*.

Lamps, or transformers feeding lamps, are connected to mains 1 and 2, and two similar transformers connected, as shown in Fig. 142, are used to supply three-phase currents to induction motors.

Problems.

1. A given transformer is rated at 5 kilowatts and is designed to take current from 1100-volt mains at a frequency of 60 cycles per second. Under these conditions the iron loss and the copper loss will be called normal.

The transformer is used at 6 kilowatts output at rated e. m. f. and frequency. Calculate copper loss in terms of normal.

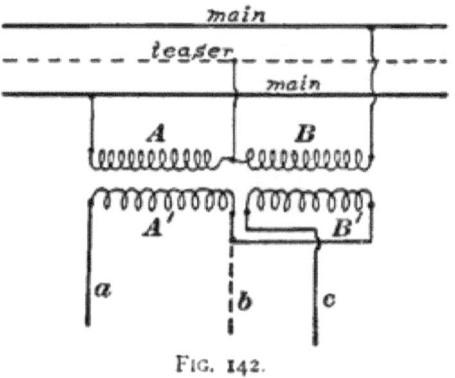

Fig. 142.

The transformer is used at rated e. m. f. but at a frequency of 75 cycles per second. Calculate iron loss in terms of normal.

The transformer is used at rated frequency but with primary e. m. f. of 1500 volts. Calculate iron loss in terms of normal.

With primary e. m. f. of 1500 volts what frequency would give normal iron loss?

With primary e. m. f. of 1500 volts what load would give normal copper loss?

2. The core of a transformer has 7200 cubic centimeters of annealed refined sheet iron of thickness .035 centimeter. The sectional area of the magnetic circuit is 144 square centimeters and the length of the magnetic circuit is 50 centimeters. The permeability of the iron is 1250. The primary coil has 500 turns and is connected to a 110-volt alternator having a frequency

of 125 cycles per second. Calculate the following quantities: (a) Maximum core flux. (b) Maximum flux density. (c) Hysteresis loss in watts. (d) Eddy current loss in watts. (e) Magnetizing current. Also calculate phase difference between primary e. m. f. and magnetizing current. Magnetizing current here means the total current in the primary coil, circuit of secondary coil being of course open.

3. The primary and secondary e. m. f. ratings of a transformer are 1100 volts and 110 volts respectively. The secondary is connected in series with 4 ohms (noninductive) to 110-volt mains and the primary is connected to a noninductive resistance of 200 ohms. Calculate current in each coil.

The primary (1100-volt coil) is in two parts which are in series for 1100 volt primary rating. These two parts are thrown in parallel, resistance being the same as before. Calculate primary current and secondary current. In this problem ignore the resistances of the transformer coils.

4. A transformer of which the primary and secondary e. m. f.'s are rated at 1100 and 110 is connected to 1000-volt mains with its secondary connected in series with the primary. Calculate the e. m. f. between the terminals of the secondary coil (a) when the secondary coil helps the primary, (b) when the secondary coil opposes the primary.

5. The primary coils of two transformers have each 560 turns of wire and they are connected to two-phase mains, the e. m. f. of each phase being 800 volts.

Calculate the turns of wire required in each of two secondary coils (one on each transformer) so that these coils when connected in series give an e. m. f. of 400 volts 30° ahead of one of the two-phase e. m. f.'s.

CHAPTER XII.

THE SYNCHRONOUS MOTOR.

115. Two alternators in series.—Consider two alternators A and B connected in series and driven to give precisely the same frequency. Let the lines A and B, Fig. 143, represent the effective e. m. f.'s of machines A and B respectively and let φ be angular lag of the e. m. f. B behind the e. m. f. A. The vector sum of A and B, namely E, is the resultant e. m. f. of the two machines and this e. m. f., according to Prob. IV., Chap. V., produces a current

$$I = \frac{E}{\sqrt{R^2 + \omega^2 L^2}} \qquad (83)$$

which lags $\theta°$ behind E where:

$$\tan \theta = \frac{\omega L}{R} \qquad (84)$$

in which R is the total resistance of the circuit, L its total inductance including the armatures of both machines, and $\omega = 2\pi f$, f being the frequency.

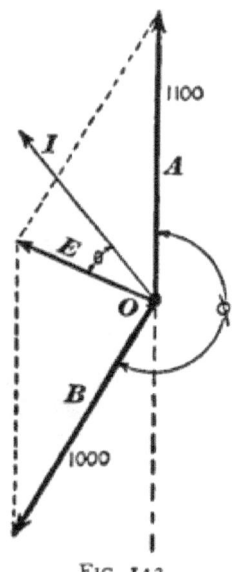

FIG. 143.

The power P' put into the circuit by machine A is

$$P' = AI \cos (AI) \qquad (85)$$

when (AI) is the angle between A and I. The power P'' put into the circuit by machine B is

$$P'' = BI \cos (BI). \qquad (86)$$

The angle (AI) in Fig. 143 is less than 90°, so that $\cos (AI)$ is positive; therefore P' is positive, that is, the machine A is acting as a dynamo. The angle (BI) in the figure is greater than 90°, so that cos (BI) is negative; therefore P'' is negative, that is, the machine B is acting as a motor.

The alternator B used in this way is called a *synchronous motor*, the alternator A being driven by an engine or water wheel.

116. Variation of P' and P'' with the Phase Angle φ.—Draw a line OC, Fig. 144, representing B to scale. Describe about C a

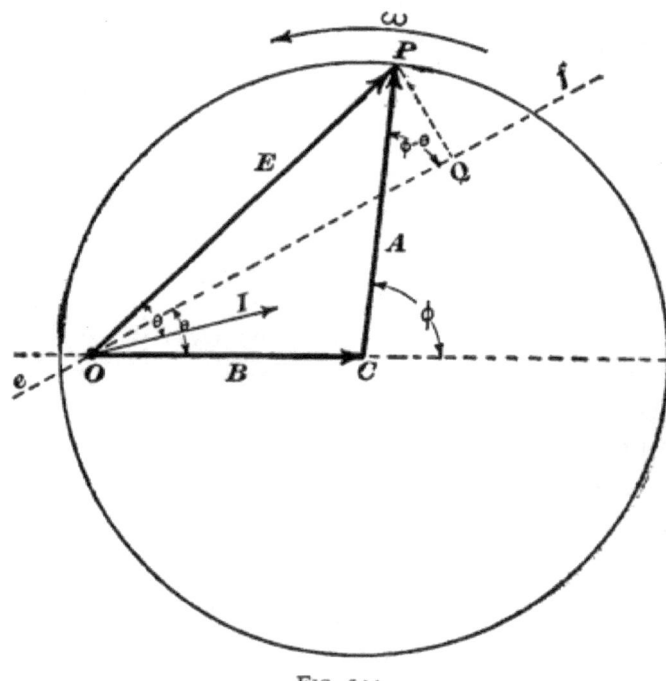

FIG. 144.

circle of which the radius represents A. Then a line OP, from O to any point in the circle represents a possible value of the resultant e. m. f. E, φ being the corresponding phase difference between A and B. Draw the line ef through O, making with OC the angle θ. Then the angle POf is equal to the angle (BI). Therefore

$$OQ = OP \cos(BI) = E \cos(BI) = \sqrt{R^2 + \omega^2 L^2}\, I \cos(BI)$$

or $I \cos(BI) = \dfrac{\overline{OQ}}{\sqrt{R^2 + \omega^2 L^2}}$. Substituting this value of $I \cos(BI)$ in equation (86) we have

$$P'' = \dfrac{B \cdot \overline{OQ}}{\sqrt{R^2 + \omega^2 L^2}} \qquad (87)$$

That is, the output of the machine B is proportional to the projection OQ of the line OP on the line of, B and $\sqrt{R^2 + \omega^2 L^2}$ being constant.

When Q is towards f from $O \cos(BI)$ is positive so that P'' is positive and machine B acts as a dynamo. When Q is towards e from $O \cos(BI)$ is negative so that P'' is negative and machine B acts as a motor.

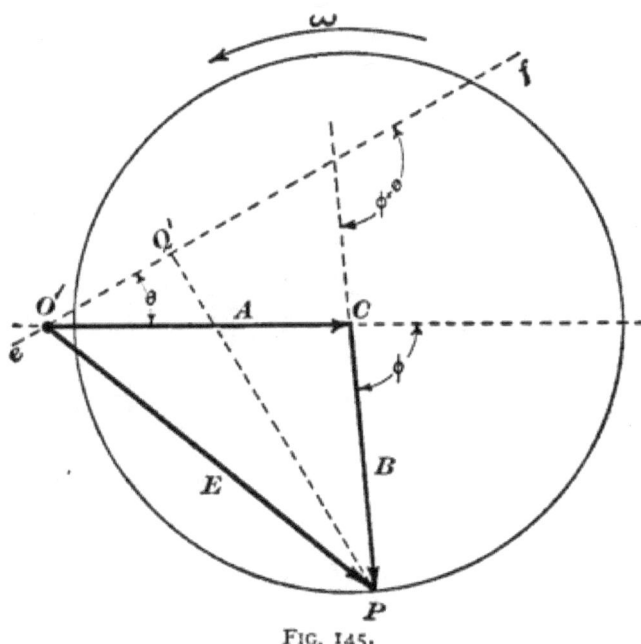

FIG. 145.

Fig. 145 is a construction, for the same value of φ, in which $O'Q'$ represents the power P' put into the circuit by machine A. From this diagram we have

$$P' = \frac{A \cdot \overline{O'Q'}}{\sqrt{R^2 + \omega^2 L^2}}. \tag{88}$$

The projection of A, Fig. 144, on ef is $A \cos(\varphi - \theta)$; the projection of B on ef is $B \cos \theta$; and OQ is the sum of these projections so that

$$OQ = A \cos(\varphi - \theta) + B \cos \theta.$$

Substituting this value of OQ in equation (87) we have

$$P'' = \frac{AB}{\sqrt{R^2 + \omega^2 L^2}} \cos(\varphi - \theta) + \frac{B^2}{\sqrt{R^2 + \omega^2 L^2}} \cos \theta. \tag{89}$$

Similarly from Fig. 145 we have

$$P' = \frac{AB}{\sqrt{R^2 + \omega^2 L^2}} \cos(\varphi + \theta) + \frac{A^2}{\sqrt{R^2 + \omega^2 L^2}} \cos \theta. \tag{90}$$

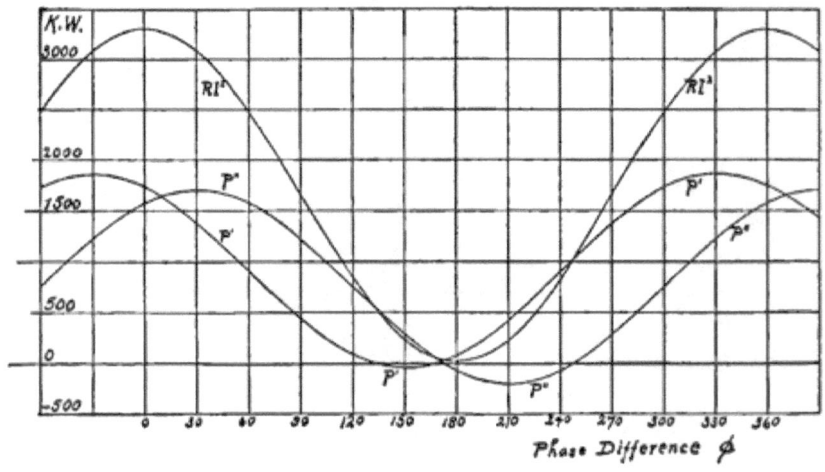

FIG. 146.

Further the algebraic sum of the outputs of machines A and B, namely, $P' + P''$ is equal to RI^2 so that

$$P' + P'' = RI^2. \tag{91}$$

These three equations, (89), (90) and (91) are the fundamental equations of the synchronous motor.

The ordinates of the curves P'', P' and RI^2, Fig. 146, show the values of P' of P'' and of RI^2 for values of φ from zero to 360°. Positive ordinates represent positive power (dynamo action), negative ordinates represent negative power (motor action). Each ordinate of the curve RI^2 is the algebraic sum of the ordinates of the curves P', P''. Fig. 147 shows portions of the curves

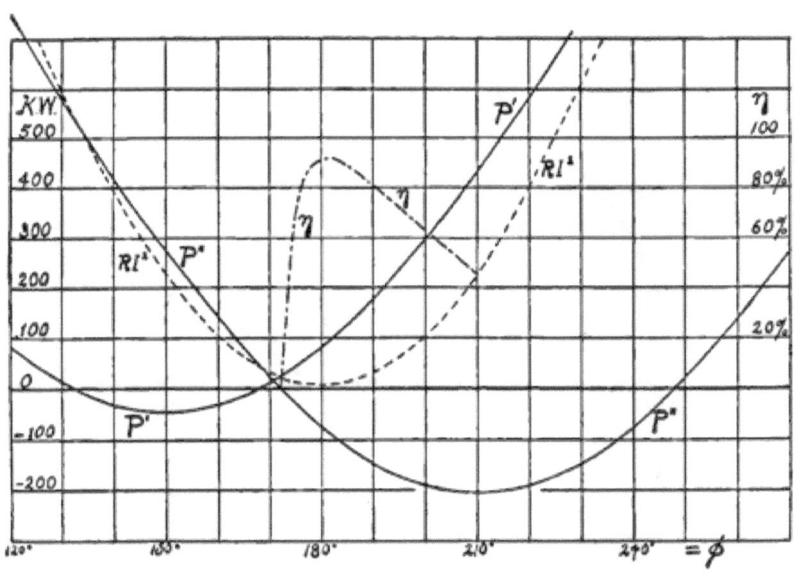

FIG. 147.

P', P'' and RI^2 to a larger scale. The ordinates of the curve η represent the efficiency $\left(\dfrac{P''}{P'}\right)$ of transmission for various values of φ when machine B is a motor.

Figs. 143 to 148 are based on the values $A = 1100$ volts, $B = 1000$ volts, $R = 1$ ohm and $\omega L = 0.58$ ohm, also the particular statements given below are based on these values.

By comparing the ordinates of the curves P' and P'', Fig. 146 or Fig. 147, it is seen that when machine A is a motor (negative values of P') its intake is very much less than the output of B. Therefore the efficiency of transmission is quite small when the

156   THE ELEMENTS OF ALTERNATING CURRENTS.

machine A, having the larger e. m. f., is the motor.* The maximum intake of machine A as a motor is 45.1 kilowatts. The maximum intake of machine B as a motor is 202.6 kilowatts.

When machine B is running as a motor, unloaded, its intake P' is approximately zero; the point P, Fig. 148, is at s,† and

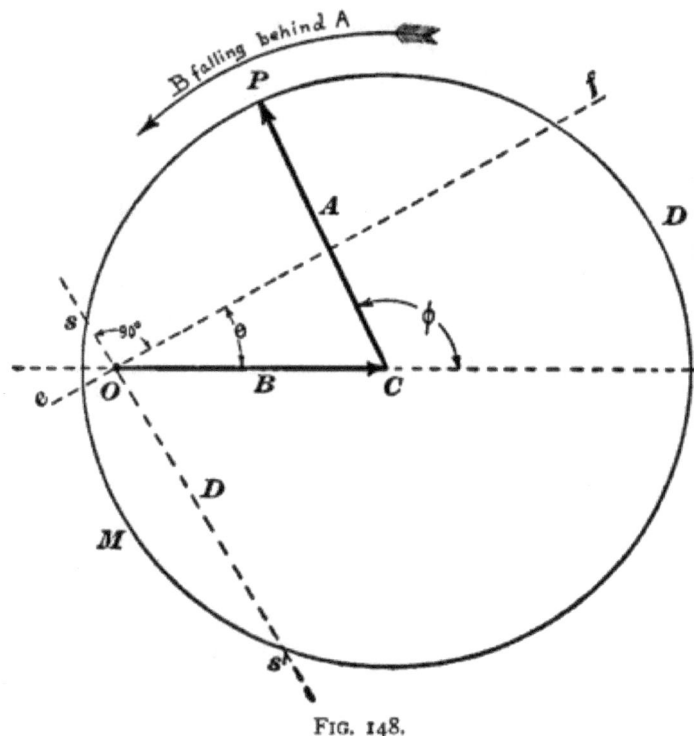

FIG. 148.

the value of φ is 171°.9. When the intake of B is zero the resultant e. m. f. is Os, Fig. 148; the current, which is in quadrature with B, is about 140 amperes; RI^2 is about 20 kilowatts; and the

---

* When the reactance ωL of the circuit is large compared with R, then the machine having the larger e. m. f. may be used as the motor without greatly reducing the efficiency of transmission; also when the reactance of the circuit is great compared with R the e. m. f. of the motor may greatly exceed the e. m. f. of the generator as will be seen later.

† Running of B is unstable when P is at S'.

output of A is of course equal to RI^2. As the motor B is loaded the point P, Fig. 148, moves from s towards M; and the resultant e. m. f., also the current and RI^2, grow less until $\varphi = 180°$. The minimum value of I is about 90 amperes, and the minimum value of RI^2 is about 8.1 kilowatts. When the point P, Fig. 148, reaches the line ef the current is opposite to the e. m. f. B in phase and the efficiency of transmission is the greatest possible for the given values of A, B, ωL and R. This maximum efficiency is about 92% and the value of φ is $182°.9$. As the motor B is still further loaded the point P moves on towards M, and when P reaches M the intake of B is at its maximum value of 202.6 kilowatts. Further loading of B causes it to fall out of synchronism and stop. In practical working the e. m. f. of the dynamo (A) is usually greater than that of the motor (B), and the load is limited to a smaller value than the maximum for the sake of high efficiency and of stability.

Remark: When the polyphase alternator is used as a synchronous motor each armature winding of the machine receives current from one phase of a polyphase system, and the total power intake is two times or three times as great as the intake of each winding, according as the machine is a two-phase or a three-phase machine. The present chapter deals explicitly with the single-phase synchronous motor, but the entire discussion applies equally well to the polyphase machine. For example: OC, Fig. 144, may represent the e. m. f. of one winding of a polyphase synchronous motor and A the e. m. f. of one phase of the supply dynamo, then P' equation (89) is one-half or one-third of the total power intake of machine B and RI^2 equation (91) is the watts lost in one circuit only of the polyphase system.

117. The starting of the synchronous motor.—To get the machine B into operation as a synchronous motor it must * be started by some independent mover, by an induction motor, for example,

---

* A polyphase synchronous motor is an induction motor when running below synchronism and is self-starting.

and its speed carefully adjusted until it is in synchronism with the machine A, which has been started up beforehand. So long as the frequency of B is less than the frequency of A the angle φ, Figs. 143 and 144, increases continuously, the point P, Fig. 144, passes once around the circle while A gains one cycle on B, and, if the circuit contains no auxiliary resistance, the outputs of A and B, namely P' and P'', pass repeatedly through the series of values shown in Fig. 146. During the time, however, that the machine B is being adjusted to synchronism with A an auxiliary resistance, usually a lamp, is connected in circuit, as shown in Fig.

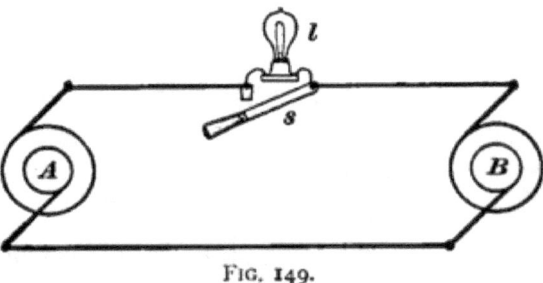

Fig. 149.

149, to limit the excessive values that would otherwise be reached by P' and P''. This lamp pulsates in brightness as machine B is being speeded up, and the pulsations become slower and slower as the frequency of B approaches the frequency of A. When the lamp is brightest, $\varphi = 0°$ or $360°$, and when the lamp is dimmest, $\varphi = 180°$, as is shown by the RI^2 curve, Fig. 146. When the machines are very nearly in synchronism the pulsations of the lamp are very slow, and the switch s is then closed as the lamp passes its minimum of brightness. The machine B is then in operation as a synchronous motor and the independent mover used to start B may be disconnected.

In practice the lamp l is connected in series with the secondaries of two transformers, the primaries of which are connected to A and to B respectively. In this arrangement the lamp may be either at its maximum or at its minimum of brightness when the proper conditions are reached for the closing of the switch s, according to the connections of the transformers.

118. Stability of running of synchronous motor.—Suppose the machine A to be driven by means of a governed engine at constant speed, irrespective of its output; and suppose the motor B to be running steadily. If the load on B is suddenly increased this machine will run momentarily slower than A and fall behind A in phase. If this falling behind in phase *increases* the power which B takes from A, then B will fall only so far behind as to enable it to take in power sufficient to carry its increased load. If, on the other hand, this falling behind in phase *decreases* the power which B takes from A, then B will fall further and further behind A, fall out of synchronism and stop. In the first case the running of B is *stable*, in the second case it is *unstable*.

As the point P, Fig. 144, moves along the circle in the direction of the arrow ω the e. m. f. A is getting farther ahead of B in phase or B is falling behind A. *If the motor intake of B increases as it falls behind A, then the running of B is stable, and vice versa*, as pointed out above. Now the projection of OP, namely OQ, positive towards c, represents the intake of B; and this intake increases from s to M, Fig. 148, as B falls behind A, and decreases from M to s' as B falls behind A; therefore s to M is the region of stable motor running of B and M to s' is the region of unstable motor running of B.

If B is running with given load as a motor the point P will take up a position between s and M such that the intake of B is sufficient to carry its load. If B is further loaded P moves further towards M; if B is unloaded P moves towards s. If B is loaded until P reaches M then further loading decreases the intake of B and the machine B therefore falls out of synchronism and stops. The action while stopping is as follows: Every time B loses one cycle as compared with A the point P moves once around the circle, Fig. 148. While P moves from s' through D to s machine B acts as a dynamo which action together with its belt load as a motor slows it up rapidly. Then as P moves from s through M to s' machine B takes in power from A but by no means enough to enable it to regain its speed, and so on.

Remark: For given intake of B (given value of OQ, Fig. 144) there are two values of the resultant e. m. f. OP and, therefore, two values of current I. The lesser value of I corresponds to stable running and the greater value to unstable running.

Running of two alternators in series as dynamos.—Suppose alternator A, Fig. 150, to be driven at constant speed by a governed engine, and alternator B to be driven by an ungoverned engine. Alternator B being once adjusted to synchronism with A keeps in synchronism and gives a constant output. The variations of total load are met by the alternator A and its governed engine. The point P, Fig. 148, stands somewhere between s and D, for in this region the output of B decreases as it falls behind and increases as it gets ahead of A so that the ungoverned engine and alternator B are in stable state of running. Alternators are never used thus in practice for the reason that the resultant e. m. f. OP, Fig. 148, varies greatly with the total load.

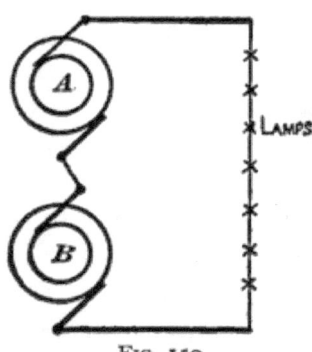

FIG. 150.

Running of two alternators in parallel as dynamos.—Two similar alternators connected in parallel, as shown in Fig. 151, run satisfactorily when they are once adjusted to synchronism and this arrangement is frequently used in practice. Machine A is started and connected through a resistance l to B; machine B is then started and when the pulsations of the lamp become very slow the

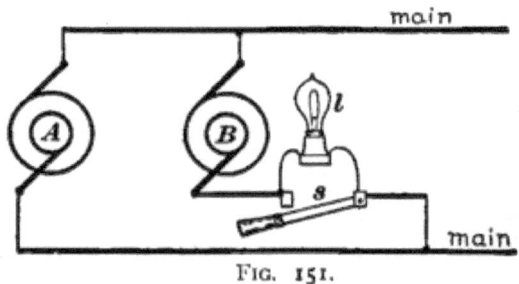

FIG. 151.

switch s is closed at minimum brightness. The two dynamos together are then connected to the mains. When two machines are run in parallel in this way the machine which pushes ahead in phase takes a larger portion of the load and the running is therefore stable. One of the alternators should be driven by an ungoverned engine.

THE SYNCHRONOUS MOTOR.

119. Greatest intake of machine B; A, B, ωL and R being given. P'' has its maximum negative value when $\cos(\varphi - \theta) = -1$ and equation (89) becomes

$$P''_{max.} = \frac{B^2 \cos \theta - AB}{\sqrt{R^2 + \omega^2 L^2}}. \tag{92}$$

Fig. 152 shows the state of affairs when intake of B is at its greatest.

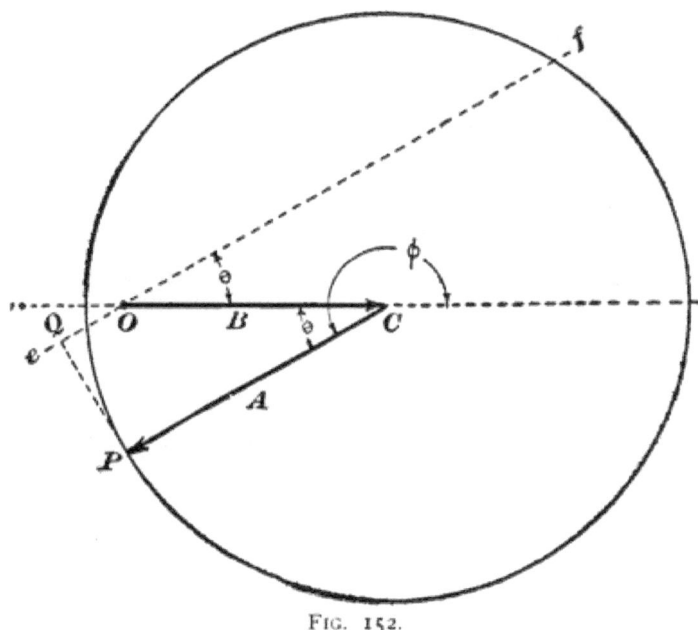

FIG. 152.

120. Greatest value of the e. m. f. B for which machine B can act as a motor; A, ωL and R being given.—So long as $\dfrac{AB}{\sqrt{R^2 + \omega^2 L^2}}$ is greater than $\dfrac{B^2}{\sqrt{R^2 + \omega^2 L^2}} \cdot \cos \theta$ then P'' can have negative values according to equation (89). Therefore the limiting case is where $\dfrac{AB}{\sqrt{R^2 + \omega^2 L^2}} = \dfrac{B^2}{\sqrt{R^2 + \omega^2 L^2}} \cdot \cos \theta$ or

$$B = \frac{A}{\cos \theta} \tag{93}$$

This limiting case is shown in Fig. 153.

121. To find value of B for which the machine B may take in the greatest possible power from A; A, ωL and R being given.—Equation (92) expresses the greatest intake of B for given values of A, B, ωL and R. It is required to find the value of B which will make this greatest intake a maximum. This value of B must render $B^2 \cos \theta - AB$ [the numerator of right hand member of equation (92)] a maximum. Differentiating this expression with respect to B and placing the differential coefficient equal to zero we have

$$A - 2B \cos \theta = 0$$

or
$$B = \frac{A}{2 \cos \theta}. \tag{94}$$

Remark: A comparison of equations (93) and (94) shows that the value of B for greatest possible intake of machine B is

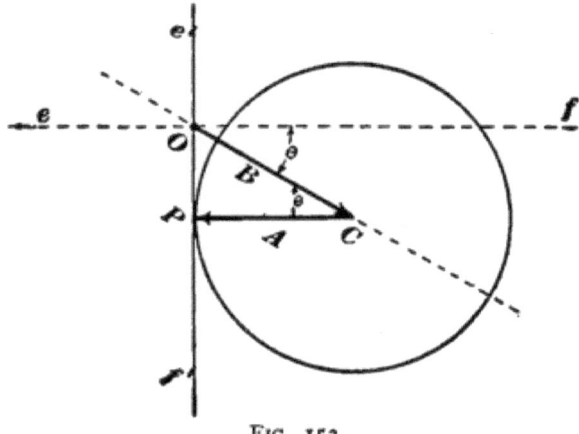

FIG. 153.

half the greatest value of B for which machine B can act as a motor at all. This is also the case with a direct-current motor. The greatest e. m. f. such a motor can have is the e. m. f. of the dynamo which drives it, and the value of its e. m. f. which permits the greatest possible intake is one-half the e. m. f. of the dynamo which drives it.

122. Value of B to give maximum intake of machine B with given current; A, ωL and R being given.—Let I, Fig. 154, be the given current and $E (= I\sqrt{R^2 + \omega^2 L^2})$ the resultant e. m. f. In order that the intake of B may be a maximum $BI \cos(BI)$ or $B \cos(BI)$ must be a maximum. Now $B \cos(BI)$ is the projection of B on the current line OI. Describe a circle, center at P, radius A; then Ox is the greatest possible value of $B \cos(BI)$ for the given current and OC is the required value of B. From the triangle OPC we have

$$B^2 = A^2 + E^2 - 2AE \cos \theta. \qquad (95)$$

123. Excitation characteristics.—With given load on a synchronous motor (given value of P'') its e. m. f. B may be changed by varying its field excitation, and for each value of B there is a definite value of the current I. Thus the abscissas of the curves, Fig. 155, represent values of I and ordinates represent values of B, for loads of zero, 100 kilowatts and 200 kilowatts respectively. These curves are called the *excitation characteristics* of the motor. Fig. 155 is based on the values $A = 1100$ volts, $R = 1$ ohm

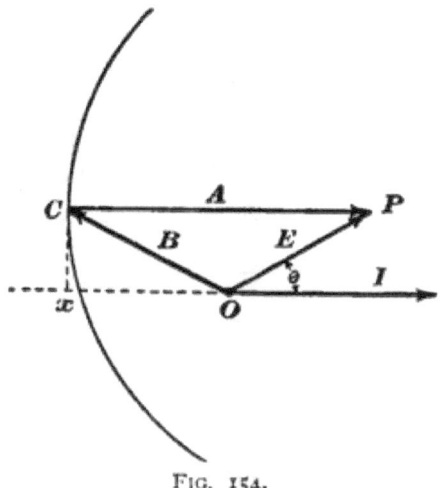

FIG. 154.

and $\omega L = 0.58$ ohm. For the greatest possible intake, 302.7 kilowatts, the characteristic reduces to the point enclosed in the small circle. It was pointed out in Art. 118 that with given load there are two values of I for each value of B and that the larger value of I corresponds to unstable and the smaller value to stable running. The dotted portions of the curves, Fig. 155, correspond to the larger values of I. These dotted portions cannot, of course, be determined by experiment on account of the instability of running.

The equation to the excitation characteristics may be derived as follows: Let I, Fig. 156, represent the current and E the resultant e. m. f.; the components of E are RI and ωLI. The e. m. f. E is the vector sum of A and B as shown and the component of B parallel to I is $\dfrac{P''}{I}$. From the right-angled triangles of the figure we have

$$B^2 = x^2 + \left(\frac{P''}{I}\right)^2 \tag{a}$$

$$A^2 = \left(RI + \frac{P''}{I}\right)^2 + (x - \omega LI)^2. \tag{b}$$

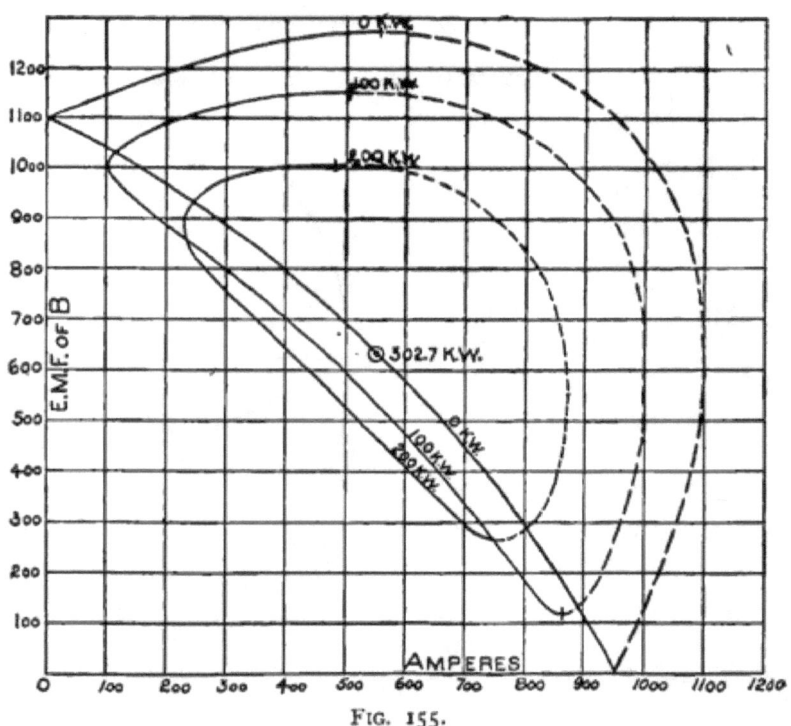

Fig. 155.

By eliminating x from these equations we have the required relation between B and I; P'', A, R and ωL being given. The curves, Fig. 155, were calculated graphically by means of the diagram, Fig. 144.

124. Value of B to bring I into phase with A; P''', A, R and ωL being given.—When I is in phase with A, the output of A is

THE SYNCHRONOUS MOTOR. 165

the greatest possible for the given current. Let OI, Fig. 157, be the current and E the resultant e. m. f.; A being parallel to I. The component of B parallel to I is $\dfrac{P''}{I}$.

From this diagram we have

$$\left(\frac{P''}{I}\right)^2 + \omega^2 L^2 I^2 = B^2,$$

and $\dfrac{P''}{I} + RI = A.$

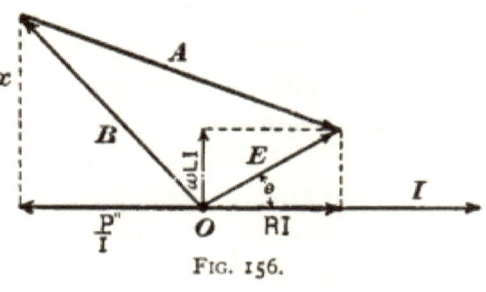

FIG. 156.

Eliminating I from these equations we have an equation which expresses the required value of B in terms of P'', A, R and ωL.

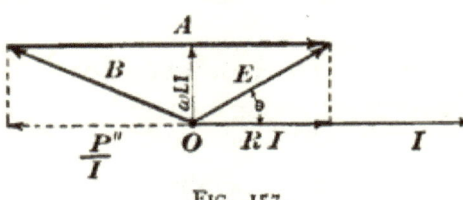

FIG. 157.

125. Value of B to bring I into phase with B; P'', A, R and ωL being given.—When I is in phase with B the intake of B is the greatest possible for the given values of B and I.

From Fig. 158 we have

$$(B + RI)^2 + \omega^2 L^2 I^2 = A^2. \qquad (a)$$

Further $\qquad P'' = BI. \qquad (b)$

Eliminating I from these equations we have an equation which expresses the required value of B in terms of P'', A, R and ωL.

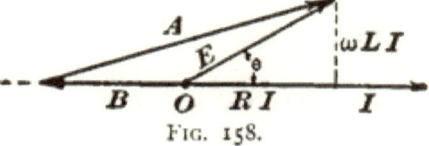

FIG. 158.

Remark: The discussion in this chapter applies to the *rotary converter* which is a synchronous motor in every essential particular.

CHAPTER XIII.

THE ROTARY CONVERTER.

126. The rotary converter.—An ordinary direct current dynamo may be made into an alternator by providing it with collecting rings, as described below, in addition to its commutator. Such a machine is called a *rotary converter*.

The single-phase converter is provided with two collecting rings which, in case of a two-pole machine, are connected to diametrically opposite armature conductors.

The two-phase converter is provided with four collecting rings which, in case of a two-pole machine, are connected to armature conductors 90° apart.

The three-phase converter is provided with three collecting rings which, in case of a two-pole machine, are connected to armature conductors 120° apart.

Remark: It is often convenient to refer to a rotary converter as a two-ring, three-ring, four-ring, or n-ring converter, as the case may be.

Remark: In case of a multipolar machine the n collecting rings are connected to the armature as follows: Ring No. 1 is connected to all armature conductors which, for any given position of the armature, lie midway under the north poles of the field magnet. Let l be the distance between adjacent conductors of this first set, that is, the distance from north pole to the next north pole. Then ring No. 2 is connected to the armature conductors which are $\frac{1}{n}$th of l ahead of the first set; ring No. 3 is connected to the armature conductors which are $\frac{2}{n}$ths of l ahead of the first set; ring No. 4 is connected to the conductors which

THE ROTARY CONVERTER. 167

are $\frac{3}{n}$ths of l ahead of the first set and so on. This statement applies to multicircuit winding. In case of the two-circuit winding each collecting ring is connected to one armature conductor only. Fig. 159 shows a four-pole dynamo with two collecting rings each connected to two armature conductors. The machine when provided with these collecting rings is a four-pole single-phase rotary converter.

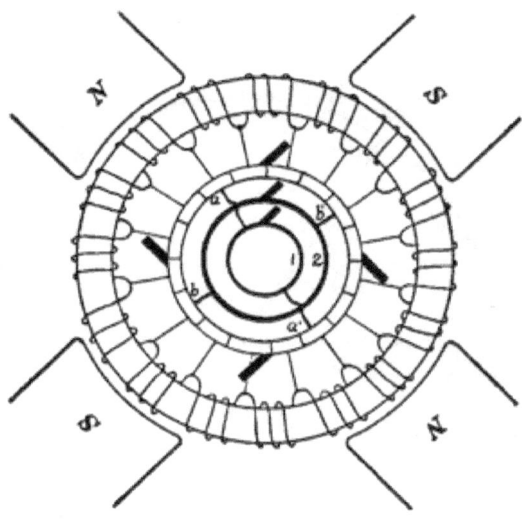

FIG. 159.

Use of the rotary converter.—The rotary converter may be used as an ordinary direct-current dynamo or motor; as an alternator or synchronous motor; it may be driven as a direct-current motor the load being provided by taking alternating current from its collecting rings; or it may be driven as a synchronous alternating current motor, the load being provided by taking direct current off the commutator. This last is the principal use of the machine. In every case in which power, transmitted to a distance by alternating current, is to be used in the form of direct current the rotary converter is used for bringing about the conversion from alternating current to direct current. Thus, in many extended

168   THE ELEMENTS OF ALTERNATING CURRENTS.

electric railway plants it is found to be expedient to transmit the power high pressure polyphase from a central station to rotary converters stationed along the line of the railway; these rotary converters in their turn supply direct current at medium pressure to the trolley wires.

127. The starting of the rotary converter and its operation when used to convert alternating current into direct current.—When used in this way the rotary converter is a synchronous motor and it differs but little in its operation from the synchronous motor with a belt load.

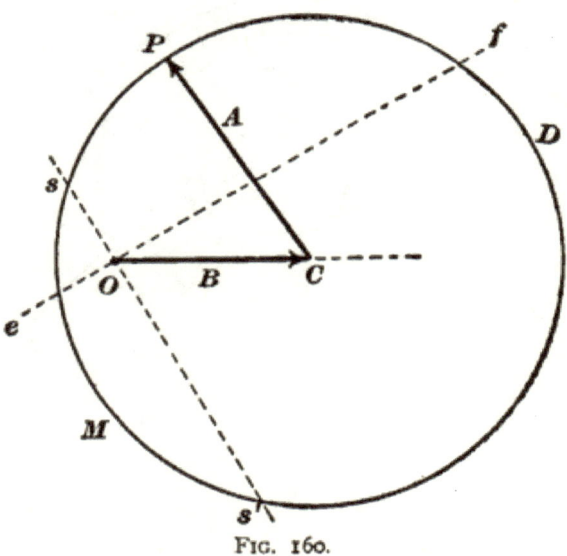

FIG. 160.

Starting.—The machine may be started as a direct-current motor using storage batteries or other local source of direct current; or it may be started in precisely the same manner as a synchronous motor with a belt load as described in Art. 117. The field magnet of a rotary converter is always excited by direct current taken from the machine itself.

Operation.—Let B, Fig. 160, be the effective alternating e. m. f. of a rotary converter and A the e. m. f. of the alternating generator. Fig. 160 is identical to Fig. 148, Chapter XII. When no direct current is taken from the converter its load is zero and the point P, Fig. 160, is at s. When direct current is taken from the converter the point P moves towards M. The alternating current taken by the converter, being proportional to the resultant e. m. f. OP, at first decreases and then increases* as the direct current load increases, and so on, exactly as in case of a synchronous motor with a belt load. See Art. 116.

It was pointed out in Chapter XII that a synchronous motor may operate at comparatively high efficiency for a wide range of values of B (value of A given) if the alternating-current circuit has large reactance; in fact, B may even be larger than A, as pointed out in Art. 120. If a considerable portion of the reactance is external to the armature of the converter then the e. m. f. between the collecting rings of the converter changes with B and so also does the direct e. m. f. of the machine. Therefore, the direct e. m. f. of a rotary converter may be varied at will† by changing the field excitation of the machine, although the e. m. f. A of the alternating generator may be constant.

128. Pumping or hunting action of synchronous motors and rotary converters.—When the load on a synchronous motor is increased the motor slows up momentarily and falls behind the generator in phase. When the motor has fallen behind sufficiently to take in power enough to enable it to carry its load it is still running slightly below synchronism; it therefore falls still further behind and takes an excess of power from the generator which quickly speeds it above synchronism. It then gains on the generator in phase until it takes in less power than is required for its load, when it again slows up and so on. This oscillation of speed above and below synchronism, called *hunting*, is similar to the hunting of a governed steam engine. It is frequently a source of great annoyance, especially where several synchronous motors or rotary converters are run in parallel from the same mains.

---

* When B is less than A.

† The possible range of variation depends upon the reactance in the circuit external to the rotary converter.

129. Armature current of a rotary converter.—Consider a given armature conductor of a rotary converter. A part of the current in this conductor is due to the alternating currents which flow into the armature at the collecting rings and a part is due to the direct current flowing out of the armature at the direct-current brushes. The actual current in the conductor is the algebraic sum of these two parts, and since these parts are generally opposite in sign, therefore the actual current in the conductor is rather small and so also is its magnetic effect and its heating effect.

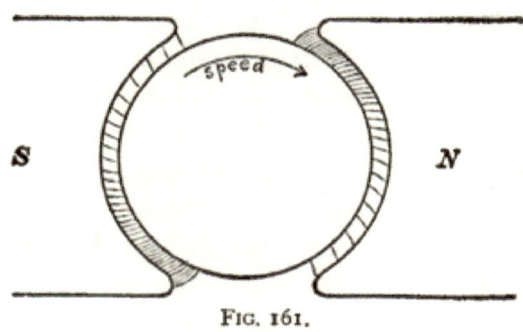

Fig. 161.

130. Magnetic reaction of the armature of the rotary converter. *Distortion of field.*—The distortion of the magnetic field of a dynamo by the armature currents accompanies, and is in fact the cause of, the torque with which the field acts upon the armature. When the torque is in the direction of the rotation of the armature (motor action) the field is concentrated under the leading horns of the pole pieces as shown in Fig. 161. When the torque is opposite to speed the field is concentrated under the trailing horns of the pole pieces as shown in Fig. 162.

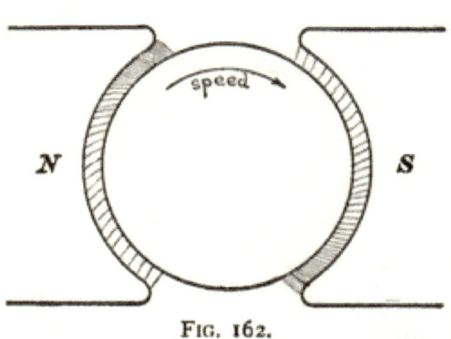

Fig. 162.

When a rotary converter is running steadily the speed of its armature is constant, and the only torque acting on the armature is the slight torque needed to overcome friction, therefore the field is scarcely at all distorted.

When a rotary converter hunts, its speed oscillates above and below synchronism so that a torque acts upon the armature, first in one direction and then in the other, and the field is concentrated, first under the trailing horns and then under the leading horns of the pole pieces.

Demagnetizing action.—The demagnetizing action of the armature currents of a rotary converter may be considered as made up of the demagnetizing action of the direct current alone and of the alternating currents alone. The first is the same as in the direct-current dynamo and the second is considered in Art. 74.

131. Power rating of rotary converters.—The magnetic action (demagnetizing action and distorting action) of the armature currents of a rotary converter is never troublesome, so that the allowable output is limited by the permissible heating of the armature. The armature heating is rather small, as pointed out in Art. 129, so that a given machine has a higher power rating as a rotary converter than as a direct-current dynamo, except in the case of the single-phase converter. The accompanying table gives the power ratings (based upon equal average armature heating) of a given machine when used (*a*) as a direct-current dynamo, (*b*) as a single-phase converter, (*c*) as a three-phase converter, (*d*) as a two-phase (four-ring) converter, and (*e*) as a six-phase converter.

POWER RATINGS OF ROTARY CONVERTERS.*

a. Continuous-current dynamo.	*b.* Single-phase converter.	*c.* Three-phase converter.	*d.* Four-ring converter.	*e.* Six-phase converter.
1.00	.85	1.32	1.62	1.92

* These ratings are calculated as explained in Art. 134, and in their calculation the losses in the machine and the wattless component of the alternating currents have been ignored. These ratings are therefore five or six per cent. too large.

132. Electromotive force relations of the rotary converter.—Let E_0 be the e. m. f. between the direct-current brushes and E_n the effective alternating e. m. f. between adjacent collecting rings of an n-ring converter. The ratio $\dfrac{E_n}{E_0}$ has a characteristic value for each value of n.

Fundamental assumption.—Consider an armature conductor c, Fig. 163, at angular distance β from the axis of the field, as

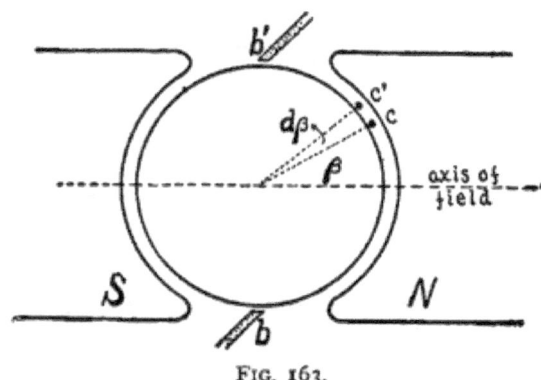

FIG. 163.

shown.* We assume that the e. m. f. induced in the conductor c is proportional to $\cos \beta$ or equal to $C \cos \beta$ where C is a constant. The results of this assumption are practically in accord with experiment.

The number of armature conductors between c and c' is proportional to, or say equal to, $d_i\beta$.

The e. m. f. in each conductor is $C \cos \beta$, and the e. m. f. in all the conductors between c and c' is:

$$de = C \cos \beta \, d\beta. \qquad (a)$$

E. m. f. E_0 between direct current brushes.—All the conductors between b and b' are in series between the direct-current brushes so that

---

* The discussion in Arts. 132, 133 and 134 is given for the case of a two-pole machine. The results, however, apply to multipolar machines as well.

$$E_0 = \int_{-90°}^{+90°} C \cos \beta \, d\beta$$

or $E_0 = 2C$. (b)

Effective e. m. f. E_n between adjacent collecting rings of an n-ring converter.—The e. m. f. between adjacent rings r and r', Fig. 164, is at its maximum value when the arc rr' is bisected by the axis

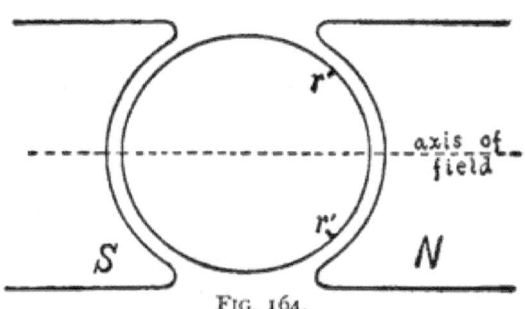

FIG. 164.

of the field as shown. The angle between r and r' is $\dfrac{2\pi}{n}$ or half this angle is $\dfrac{\pi}{n}$. The maximum e. m. f. $\sqrt{2}\,E_n$ between rings r and r' is therefore:

$$\sqrt{2}\,E_n = \int_{-\frac{\pi}{n}}^{+\frac{\pi}{n}} C \cos \beta \, d\beta = 2C \sin \frac{\pi}{n}$$

or since $2C = E_0$ we have.

$$E_n = \frac{1}{\sqrt{2}} E_0 \sin \pi/n. \tag{96}$$

Examples: The effective alternating e. m. f. of a single-phase converter ($n = 2$) is:

$$E_2 = \frac{E_0}{\sqrt{2}} = .707\,E_0. \tag{97}$$

The effective alternating e. m. f. between adjacent rings of a three-phase converter ($n = 3$) is:

$$E_3 = \frac{\sqrt{3}\,E_0}{2\sqrt{2}} = .612\,E_0. \tag{98}$$

The effective alternating e. m. f. between adjacent rings of a two-phase converter ($n = 4$) is:

$$E_4 = \frac{E_0}{2}. \qquad (99)$$

The effective e. m. f. between opposite rings of a two-phase converter is E_2.

133. Current relations of the rotary converter. *Fundamental assumptions.*—In the discussions of current relations we shall assume that the alternating current flowing through each section (between adjacent collecting rings) of the armature is exactly opposite in phase to the alternating e. m. f. in that section, and that the intake and output of power are equal.

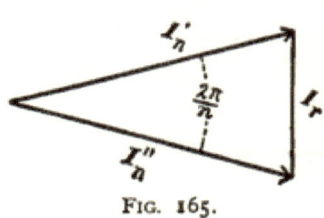

Fig. 165.

Let I_0 be the output of direct current and let I_n be the effective alternating current flowing in the armature between two adjacent collecting rings. The intake of power per phase is $E_n I_n$ or the total intake is $n E_n I_n$ and the power output is $E_0 I_0$. Therefore

$$E_0 I_0 = n E_n I_n$$

or substituting the value of E_n from equation (96) we have:

$$I_n = \frac{\sqrt{2} I_0}{n \sin \pi/n}. \qquad (100)$$

Current in each main.—The current I_r in each main or the current entering the armature at each collecting ring, is the vector difference between I_n' and I_n'', Fig. 165, so that

$$I_r = 2 I_n \sin \pi/n. \qquad (101)$$

Examples: The effective alternating current in each half of the armature of a single-phase converter ($n = 2$) is:

$$I_2 = \frac{I_0}{\sqrt{2}} \qquad (102)$$

and the effective alternating current entering at each collecting ring is
$$I_r = 2I_2 = 1.414 I_0. \quad (103)$$

The effective alternating current flowing in the armature between adjacent collecting rings of a three-phase converter ($n = 3$) is:
$$I_3 = \frac{2\sqrt{2}}{3\sqrt{3}} I_0 \quad (104)$$

and the effective current entering at each collecting ring is
$$I_r = \sqrt{3} I_3. \quad (105)$$

The effective alternating current flowing in the armature between adjacent collecting rings of a two-phase converter ($n = 4$) is:
$$I_4 = \tfrac{1}{2} I_0 \quad (106)$$

and the effective current entering at each collecting ring is
$$I_r = \sqrt{2} I_4. \quad (107)$$

134. Discussion of armature heating of the rotary converter.—Let r and r', Fig. 166, be the points of attachment of adjacent

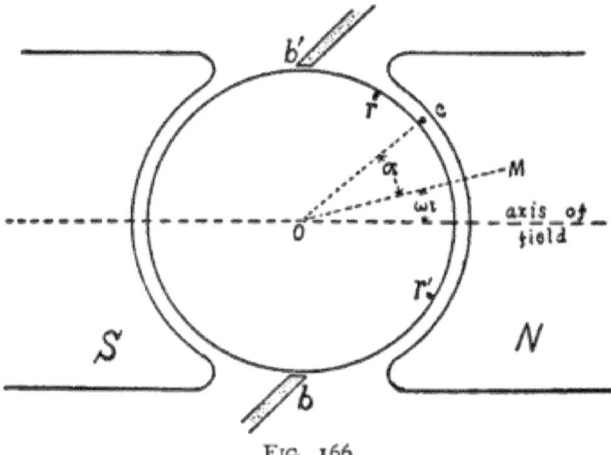

FIG. 166.

collecting rings of an n-ring converter and let the line OM bisect the arc $r\,r'$. Consider an armature conductor c between r and r' and let the angle cOM be represented by a. The largest possible value of a is π/n or one-half of the angle between r and r'.

Let ωt be the angle between OM and the axis of the field. The conductor c is at the brush b when $\omega t = -(90° + a)$, and at the brush b' when $\omega t = 90° - a$. During this time half of the direct current I_0 flows through the conductor c in an unchanging direction.

The alternating current I_n (effective value) flowing between rings r and r' is at its maximum value $\sqrt{2}\,I_n$ when the angle ωt, Fig. 166, is zero.* This alternating current is therefore equal to $\sqrt{2}\,I_n \cos \omega t$ at each instant, and it is opposite to the direct current in direction. Therefore the total current in the conductor c during the time from $\omega t = -(90° + a)$ to $\omega t = 90° - a$, is:

$$i = \frac{I_0}{2} - \sqrt{2}\,I_n \cos \omega t,$$

or, using the value of I_n from equation (100), we have:

$$i = \frac{I_0}{2}\left(1 - \frac{4}{n \sin \pi/n}\right). \tag{108}$$

The average square of this current during the time $\omega t = -(90° + a)$ to $\omega t = 90° - a$ is:

$$\frac{I_0^2}{4}\left(1 - \frac{16 \cos a}{\pi n \sin \pi/n} + \frac{8}{n^2 \sin^2 \pi/n}\right), \tag{a}$$

and the heat generated in the conductor c is proportional to this average square. The heat generated in the conductor by the direct current alone is proportional to $\frac{I_0^2}{4}$. Therefore the conductor c has $\left(1 - \frac{16 \cos a}{\pi n \sin \pi/n} + \frac{8}{n^2 \sin^2 \pi/n}\right)$† times as much heat generated in it when the machine is used as an n-ring converter as is generated in it when the machine gives the same output of direct current as a dynamo.

---

* I_n and E_n assumed to be exactly opposite in phase; see Art. 133.
† Generally less than unity.

THE ROTARY CONVERTER. 177

The average heating over the entire armature is found by integrating the expression (a) with respect to α from $\alpha = -\dfrac{\pi}{n}$ to $\alpha = +\dfrac{\pi}{n}$ and dividing the result by $\dfrac{2\pi}{n}$. This gives:

Average heating of armature of n-ring converter is proportional to:
$$\dfrac{I_0^2}{4}\left(1 - \dfrac{16}{\pi^2} + \dfrac{8}{n^2 \sin^2 \pi/n}\right). \quad (b)$$

This average heating is therefore $\left(1 - \dfrac{16}{\pi^2} + \dfrac{8}{n^2 \sin^2 \pi/n}\right)^*$ times as great as the heating of the armature by the direct current alone. Therefore an n-ring converter can put out

$$\dfrac{1}{\sqrt{1 - \dfrac{16}{\pi^2} + \dfrac{8}{n^2 \sin^2 \pi/n}}}$$

times as much direct current as the same machine can when used as a simple dynamo, for the same total armature heating. The table given in Art. 131 is calculated in this way.

Remark: The conductors midway between the points of attachment of the collecting rings ($\alpha = 0$) are heated least, and the conductors near the points of attachment of the collecting rings $\left(\alpha = \pm \dfrac{\pi}{n}\right)$ are heated most. Thus in a two-ring converter ($n = 2$) the conductors midway between the points of attachment have, according to the expression (a), $\tfrac{45}{100}$ as much heat generated in them as would be generated by the direct current alone; and the conductors near the points of attachment of the collecting rings have 3 times as much heat generated in them as would be generated by the direct current alone.

---

* Generally less than unity.

CHAPTER XIV.

THE INDUCTION MOTOR.

135. The utilization of alternating current for motive purposes.—It has been already pointed out that the successful employment of alternating current for motive purposes depends upon the use of the *induction motor* driven by polyphase currents. The synchronous motor (and rotary converter) operates satisfactorily with single-phase current when once it is started, but if power is to be taken directly from the supply mains for starting, then polyphase currents are most satisfactory, inasmuch as the synchronous motor may then be started by an auxiliary induction motor.*

136. The induction motor consists of a primary member and a secondary member, each with a winding of wire. The primary

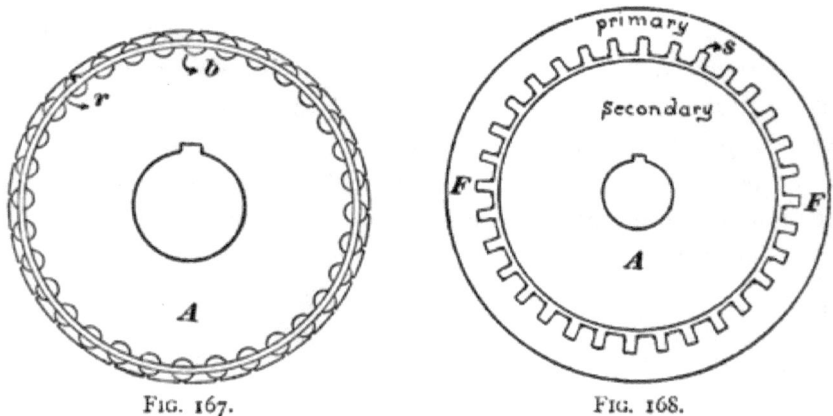

FIG. 167. FIG. 168.

member is usually stationary, and is often called the *stator*. The

---

* The polyphase synchronous motor itself acts as an induction motor when running below synchronism, especially if its field coils are short-circuited, and it may, therefore, be started without the use of an auxiliary induction motor. This method of starting is used by the General Electric Company.

secondary member is usually the rotating member, and is often called the *rotor*. Fig. 167 shows a rotor of the *squirrel cage* type. It consists of a drum, *A*, built up of circular sheet-iron disks; near the periphery of this drum are a number of holes parallel to the axis of the drum; in these holes heavy copper rods, *b*, are placed, and the projecting ends of these rods are soldered to massive copper rings, *r*, one at each end of the drum. Another type of rotor is described later.

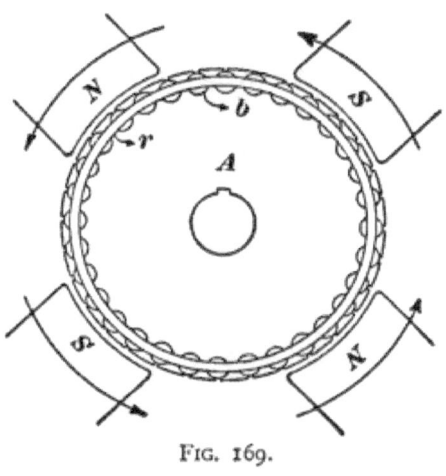

Fig. 169.

The stator is a laminated iron ring, *FF*, Fig. 168, closely surrounding the rotor. This ring is slotted on its inner face, as shown, windings are arranged in these slots, and these windings receive currents from polyphase supply mains. These polyphase currents produce in the stator a rotating state of magnetism, the action of which on the rotor is the same as the action of an ordinary field magnet in rotation. Thus, Fig. 169 shows a squirrel cage rotor, *A*, surrounded by an ordinary field magnet rotating in the direction of the curved arrows. This motion of the field magnet induces currents in the short-circuited copper rods of the rotor; the field magnet exerts a dragging force on these currents and causes the rotor to rotate.

No electrical connections of any kind are made to the rotor. The next article describes the stator windings and explains the manner in which these windings produce the rotating state of magnetism in the stator.

137. Stator windings and their action.—The stator windings are arranged in the slots *s*, Fig. 168, in a manner exactly similar

180   THE ELEMENTS OF ALTERNATING CURRENTS.

to the arrangement of the windings of the two-phase or three-phase alternator armature, according as the motor is to be supplied with two or three-phase currents.

Fig. 170 shows an end view of a four-pole two-phase induction motor. In this figure the outline, only, of the rotor is shown; the stator conductors are represented in section by the small circles; the slots are omitted for the sake of clearness; and the end connections of half the stator conductors are shown in Fig. 171. The stator conductors are arranged in two distinct circuits. One of these circuits includes all of the conductors marked A and receives current from one phase of a two-phase system; the other circuit includes all of the conductors marked B and receives current from the other phase of the two-phase system. The terminals of the B circuit are shown at tt', Fig. 171. The conductors which constitute one circuit are so connected that the current flows in opposite directions in adjacent groups of conductors as indicated by the arrows in Fig. 171. The radial lines in Fig. 171 represent the stator conductors and the curved lines represent the end connections, as in the winding diagrams, Figs. 98 to 105.

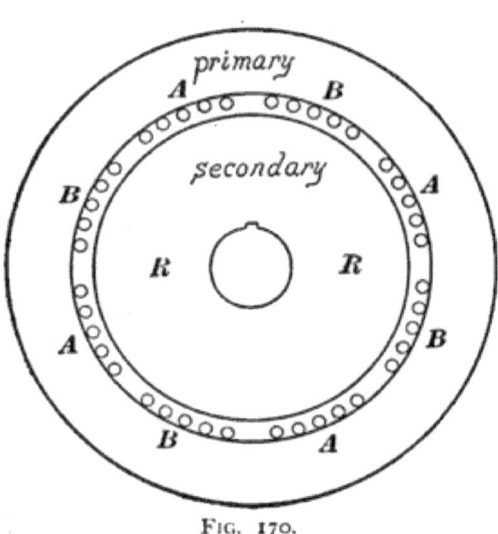

FIG. 170.

The action of a band of conductors between two masses of iron is shown in Figs. 172 and 173. The small circles in these figures represent the conductors in section; conductors carrying down-flowing currents are marked with crosses, those carrying up-flowing currents are marked with dots, and those carrying no current are left blank. The action of the currents in these bands of conductors is to produce magnetic flux along the dotted lines in the directions of the arrows.

The lines A' and B' in Figs. 174, 175 and 176 are supposed

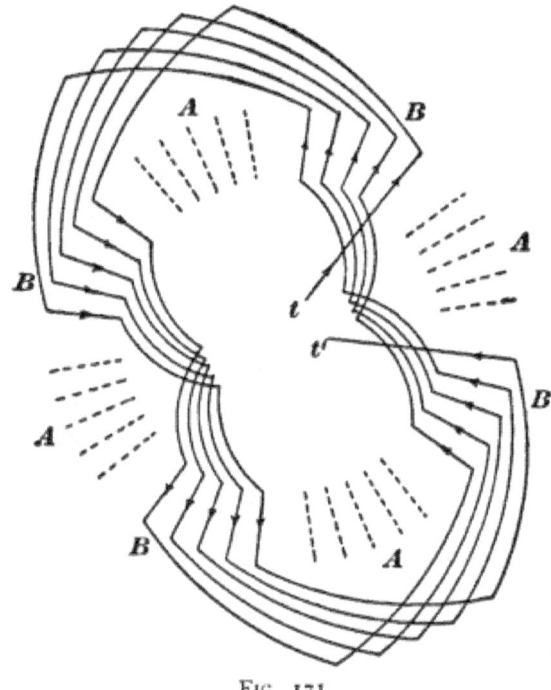

Fig. 171.

to rotate and their projections on the fixed line cf represent the instantaneous values of the alternating currents in the A and B conductors respectively.

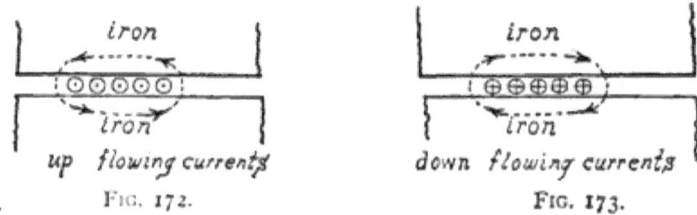

Fig. 172.   Fig. 173.

Fig. 174 shows the state of affairs when the current in conductors A is a maximum and the current in conductors B is zero.

182   THE ELEMENTS OF ALTERNATING CURRENTS.

The dotted lines indicate the trend of the magnetic flux. This flux enters the rotor from the stator at the points marked N and leaves the rotor at the points marked S.

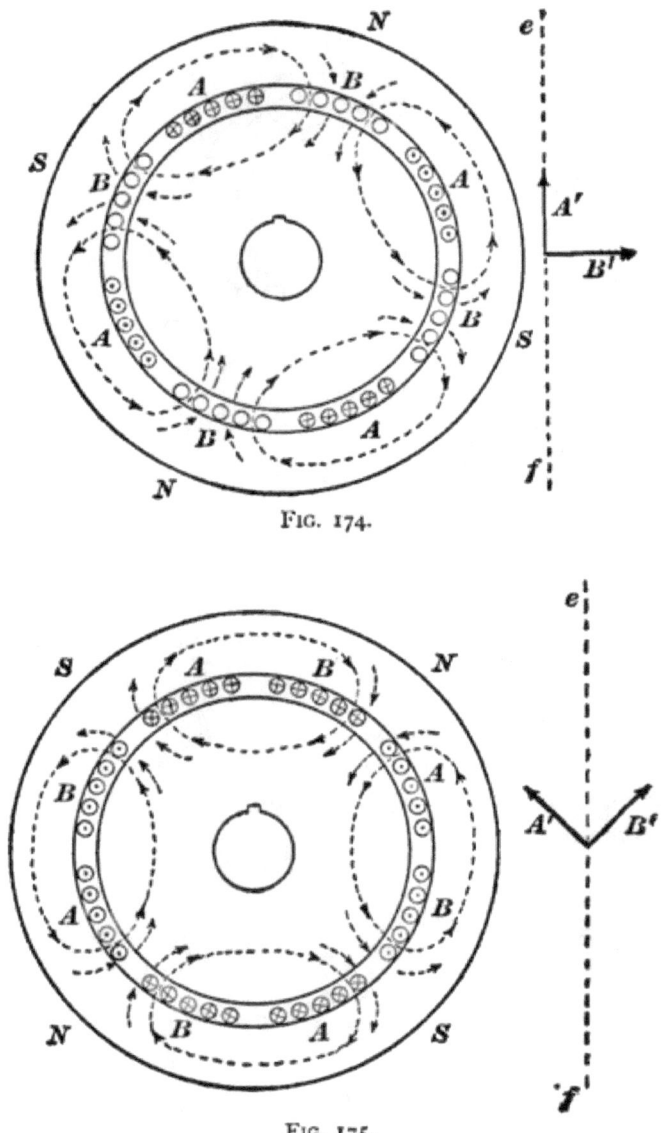

Fig. 174.

Fig. 175.

Fig. 175 shows the state of affairs, ⅛ of a cycle later, when the current in the B conductors has increased and the current in the A conductors has decreased to the same value. The points N and S have moved over $\frac{1}{16}$ of the circumference of the stator ring.

Fig. 176 shows the state of affairs, after another eighth of a cycle, when the current in the B conductors has reached its

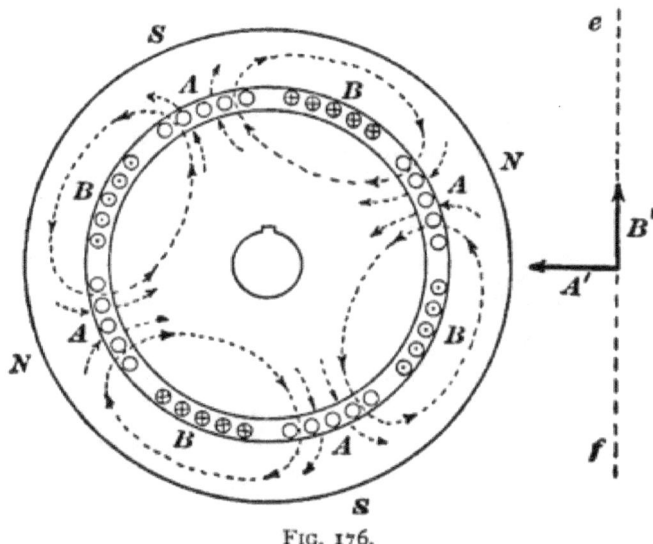

FIG. 176.

maximum value and the current in the A conductors has dropped to zero. The points N and S have moved again over $\frac{1}{16}$ of the circumference of the stator ring.

This motion of the points N and S is continuous, and these points make one complete revolution (in a four-pole motor) during two complete revolutions of the vectors A' and B' or while the alternating currents supplied to the stator windings are passing through two cycles. In general

$$n = \frac{f}{p}$$

in which n is the revolutions per second of the stator-magnetism, p is the number of pairs of poles N and S, and f is the frequency of the alternating currents supplied.

138. Preliminary discussion of the action of the induction motor.—The complete theoretical discussion of the action of the induction motor is given later and is in many respects similar to the theory of the transformer. Many important details of the action of the induction motor, however, are most easily explained by looking upon the induction motor as a rotor influenced by a rotating field magnet.

Torque and speed.—Let n be the revolutions per second of the field and n' the revolutions per second of the rotor. When $n = n'$ the rotor and field turn at the same speed, so that their relative motion is zero; no e. m. f. is then induced in the rotor conductors and no current, and therefore the rotating field exerts no torque upon the rotor. As the speed of the rotor decreases the relative speed of rotor and field increases, and therefore the e. m. f. induced in the rotor conductors, the currents in the conductors, and the torque with which the field drags the rotor, all increase. If the whole of the field flux were to pass into the rotor and out again in spite of the demagnetizing action of the currents in the rotor conductors, then the torque would increase in strict proportion to $n - n'$, but in fact a larger and larger portion of the field flux passes through the space between stator and rotor conductors as the speed of the rotor decreases and this magnetic leakage causes the torque to increase more and more slowly as $n - n'$ increases, only in some cases* reaching a maximum value and then decreasing with further incease of $n - n'$.

Fig. 177 shows the typical relation between torque and speed of an induction motor. Ordinates of the curve represent torque and abscissas measured from O represent rotor speeds. The rotor is said to run above synchronism when it is driven so that $n' > n$.

Use of starting resistance in the rotor windings.—The speed of

---

* In every case, if one makes $n - n'$ large enough by driving the rotor backwards so that n' becomes negative.

rotor for which the maximum torque occurs depends upon the resistance of the rotor windings, and it is advantageous to provide at starting such resistance in these windings as to produce the maximum torque at once, this resistance being cut out as the motor approaches full speed.

Efficiency and speed.—Let T be the torque with which the rotating field drags on the rotor; then $2\pi n'T$ is the power taken up by the rotor to be given off its belt pulley. Also T is the reacting torque which opposes the rotation of the field, so that $2\pi nT$

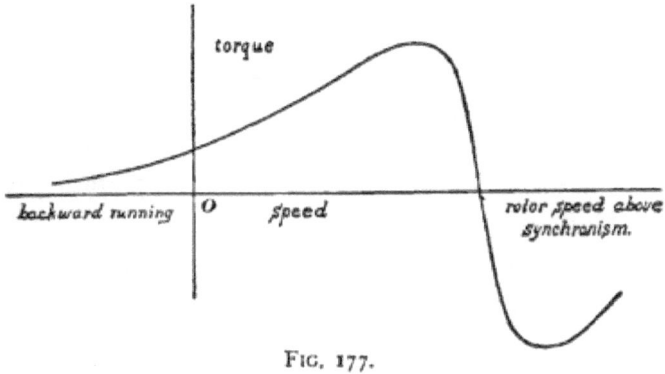

FIG. 177.

is the power required to maintain the rotation of the field. Therefore, ignoring friction losses, $2\pi nT$ is the power intake and $2\pi n'T$ is the power output of the motor, so that:

$$\eta = \frac{2\pi n'T}{2\pi nT} = \frac{n'}{n} \qquad (109)$$

is the efficiency of the machine. This equation shows that the efficiency of an induction motor is zero when the rotor stands still, that it increases as the rotor speeds up, and approaches 100% (ignoring field losses and friction) as the rotor speed approaches the field speed. The ratio $\frac{n'}{n}$ ranges from .85 to .95 or more in commercial induction motors under full load.

Efficiency and rotor resistance.—For a given difference $n - n'$ between field speed and rotor speed, given e. m. f. is induced in the rotor conductors, and the less the rotor resistance the greater the current produced by this e. m. f. and the greater the torque. Therefore a given induction motor will develop its full load torque for a small value of $n - n'$ or for a larger value of $\dfrac{n'}{n}$ (efficiency) if its rotor resistance is small. High efficiency depends, therefore, upon low rotor resistance.

The induction generator.—When the rotor of an induction motor is driven above synchronism ($n' > n$), by an engine for example, the torque is reversed and opposes the motion of the rotor so that $2\pi n'T$ is input and $2\pi nT$ is output. That is, the machine takes power from the engine to drive its rotor and gives out power from its stator windings. This output of power is in the form of polyphase currents the frequency of which is fixed by the frequency of the alternator (or synchronous motor) which is connected to the stator windings.

139. The driving of induction motors by single-phase alternating current.—This is accomplished by connecting the two stator circuits A and B (case of two-phase motor) in parallel to the single-phase supply mains at the same time connecting in series with A a resistance R (see Fig. 178). The currents in the circuits A and B then differ in phase on account of the dissimilarity of the circuits, and the motor starts. When the motor is well under way one of the windings A or B may be short-circuited, the other only being left connected to the mains. A two-phase, or even a three-phase motor operates fairly well under these conditions, except that excessive current is required at starting to give a good starting torque. The resistance R may be replaced with advantage by a condenser, especially in case of a small motor.

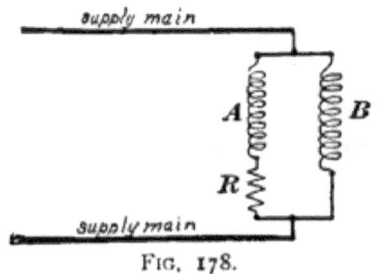

Fig. 178.

Remark: The foregoing discussion refers explicitly to the two phase motor. The three-phase motor differs but little from the two-phase motor, as will appear in the following discussion.

GENERAL THEORY OF THE INDUCTION MOTOR.

140. The general alternating current transformer.—The general theory of the induction motor is best developed by considering at once the most general type of machine, a multipolar multiphase motor, of which the rotor is wound in precisely the same way as the stator, the rotor windings being connected to collecting rings, so that the currents induced in the rotor windings may be available for outside purposes. Such a machine we will call the general alternating current transformer. Thus, a $2p$-pole, q-phase machine would have its stator conductors arranged in q distinct circuits, each taking current from one phase of a q-phase system; furthermore, each circuit would include $2p$ equidistant groups of conductors so connected that a current in that circuit would flow in opposite directions in adjacent groups. The rotor conductors would be similarly arranged in q* distinct circuits, each connected to a pair of collecting rings and supplying current to an outside receiving circuit.

Of course, no such induction motors † are ever actually built, but it is important to have clearly in mind the details of the machine to which the following discussion applies.

Fig. 179 shows a little more than one-sixth part of the circumference of a six-pole, three-phase machine. The three groups of

---

* Stator and rotor are not necessarily wound for the same number of phases, but the discussion is simplified by such an arrangement.

† Steinmetz has proposed the use of such induction motors for street railway work. Two similar motors are used on each car, one geared to each axle. At starting and for slow running motor No. 1 takes currents into its stator windings from polyphase trolley wires, and supplies polyphase currents from its rotor windings to the stator of motor No. 2, the starting resistance being connected in the rotor circuits of motor No. 2. With such an arrangement the limit of speed is one-half of synchronous speed ($n' = \frac{1}{2} n$), and the efficiency at given speed is doubled. For fast running both motors take current directly from the trolley wires.

stator conductors, A, B and C, belong one to each of the three circuits formed by the stator windings; and the three groups of rotor conductors, A', B' and C', belong one to each of the three circuits formed by the rotor windings.

When the rotor of such a machine is stationary the machine acts simply as a transformer taking q-phase currents from the supply mains into its stator windings and giving out q-phase currents of the same frequency from its rotor windings. If stator and rotor have the same number of conductors, which is understood to be the case in the following discussion, then the ratio of transformation is 1 : 1 when the rotor is stationary.

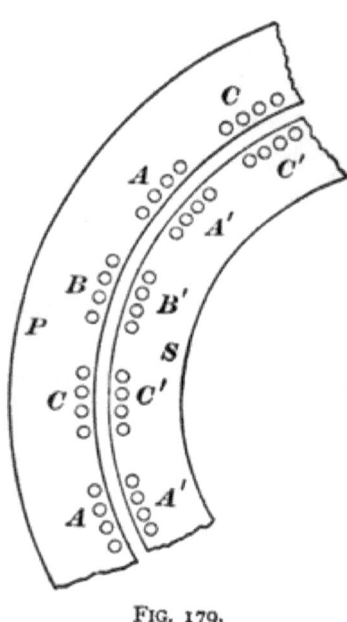

FIG. 179.

141. Effect of rotor speed upon the transformer action of the induction motor.—In the discussion of this matter it is convenient to express speed of rotor and speed of stator-magnetism not in revolutions per second but in terms of what we may call briefly *rotation frequency* which is the product of revolutions per second into the number of pairs of poles for which the machine is wound. The rotation frequency of stator magnetism is the frequency of the polyphase currents supplied to the stator windings.

Let f be the rotation frequency of stator-magnetism, and f' the rotation frequency of the rotor. The relative speed of the two is then $f - f'$.

When the rotor is stationary ($f' = 0$) the relative speed of rotor and stator-magnetism is f and the e. m. f. induced in each rotor circuit is of the same value and frequency as the e. m. f. acting on each circuit of the stator winding as pointed out in the foregoing article.

When the rotor speeds up to a rotation frequency of f' the relative speed of rotor and stator-magnetism is reduced to $f-f'$, and the e. m. f. induced in each rotor circuit is reduced *in value and in frequency* in proportion to the decrease in relative speed; that is in the ratio of $f:f-f'$. This ratio $\frac{f-f'}{f}$ is called the *slip*, s of the machine. Therefore when the rotation frequency of the rotor is f' the e. m. f. induced in each rotor circuit is sE_1 and its frequency is sf; where E_1 is the value of the e. m. f. acting on each circuit of the stator winding, stator and rotor conductors being equal in number.

The frequency changer.—The induction motor is sometimes used as a so-called frequency changer. Alternating currents at a given frequency are taken into the stator windings. The rotor is loaded by belt to bring it to such speed that currents of required frequency may be taken from the rotor windings.

142. The use of a fictitious frequency for the alternating currents in the rotor windings.—Neither the graphical method nor the symbolic method can be satisfactorily used in the discussion of an alternating current problem in which it is necessary to consider alternating currents of different frequency simultaneously, such as the alternating currents in the stator and rotor windings of the induction motor.

The e. m. f. induced in a given rotor conductor and the current flowing in the conductor actually pass through $f-f'$ cycles per second. Let us consider however, not the successive instantaneous values of e. m. f. and current in a given rotor conductor, but the instantaneous values of e. m. f. and current in the successive rotor conductors as they pass a given stator conductor. These e. m. f. and current values pass through f cycles per second, their maximum (and effective) values are the same as the maximum (and effective) values of e. m. f. and current in a given rotor conductor, and their average product gives the average power developed in a given rotor conductor.* By employing this fictitious frequency the electric and magnetic actions of the induction motor become identical to those of the simple transformer. The primary and secondary e. m. f.'s (each phase) are E_1 and sE_1 and the primary and secondary currents are equal and opposed to each other in their magnetizing action.

143. The ideal induction motor.—An induction motor of which stator and rotor† windings have negligible resistance, of which the magnetic circuits have negligible reluctance, in which hysteresis and eddy currents are negligible, and which satisfies the condition that all the magnetic flux through the stator passes also through the rotor is called an ideal induction motor.

Let E_1, Fig. 180, represent the e. m. f. acting on one circuit of the stator winding. Then $E_2\ (= sE_1)$ is the e. m. f. induced in one of the rotor circuits. The current I_2 which the e. m. f. E_2 produces is determined by the resistance and inductance of the outside circuit to which it supplies current, according to Problem IV., Chapter V. The current I_1 in the given stator circuit is equal to I_2. The input of power per phase (i. e., the power put into one stator circuit) is $E_1 I_1 \cos\theta$; the output of electrical power per phase is $E_2 I_2 \cos\theta$ or $sE_1 I_1 \cos\theta$ since $E_2 = sE_1$ and $I_2 = I_1$; and the output of mechanical power per phase is the difference between input and electrical output or $(1-s) E_1 I_1 \cos\theta$.

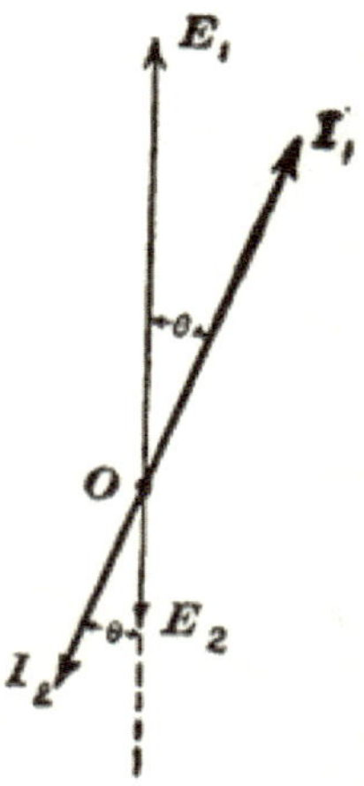

Example: A four-pole, three-phase machine takes current at 220 volts (each phase), the frequency being 60 cycles per second. When the rotor is stationary the rotor gives out three-phase currents at 220 volts and full frequency. When $f' = 30$ (rotor speed 15 revolutions per second) the rotor gives out three-phase cur-

Fig. 180.

---

* Strictly, the average power developed in all the rotor conductors while they are passing a given stator conductor—which amounts to the same thing.

† The resistance of the rotor windings may be, and is hereafter, included with the resistances of the external circuits receiving currents from the rotor.

rents at 110 volts and half frequency, and of the total power intake half is given out as mechanical energy and half as electrical energy. When $f' = 50$ (rotor speed 25 revolutions per second) the rotor gives out three-phase currents at $\frac{1}{6}$ full e. m. f. and frequency, and of the total power intake $\frac{5}{6}$ is given out as mechanical energy and $\frac{1}{6}$ as electrical energy.

When f' approaches 60 per second the e. m. f. induced in each rotor circuit approaches zero frequency and zero value, and the total intake falls off, but a larger and larger portion of the intake appears as output of mechanical energy. In this example the resistance of each circuit receiving current from rotor windings is supposed to be constant.

Remark: The action of the actual induction motor deviates from the above described ideal action because of the resistance of the stator windings, because of magnetic reluctance, eddy currents and hysteresis, and because of magnetic leakage. These things are very nearly independent of each other in their action, and therefore the effect of each will be considered by itself. Further, the discussion of the effects of these things is almost identical to the discussion of their effects on the simple transformer; therefore, those things only will be fully discussed which have a bearing upon the motor action of the machine, that is, which have influence upon the relation between torque and speed.

Magnetic leakage and rotor resistance (including resistance of entire secondary circuits) have very great influence upon the behavior of the machine as a motor; that is, upon speed and torque.

Stator resistance, magnetic reluctance, eddy currents and hysteresis are, under practical conditions, almost without influence upon speed and torque, their effect being mainly to cause the stator to take from the mains slightly more current and slightly more power for a given rotor output than would be the case with an ideal machine.

144. Effect of magnetic reluctance, eddy currents and hysteresis upon the action of an induction motor.—The ideal induction motor takes no current from the mains into its stator windings when the motor is running at synchronous speed ($f' = f$). The actual induction motor running in synchronism takes sufficient current to overcome the magnetic reluctance of the iron (stator and rotor); and an amount of power equal to the hysteresis and eddy current loss in the stator iron, only, inasmuch as the magnetic state of the rotor is constant, since it rotates with the stator magnetism. When the actual induction motor is running at any given speed it takes from the mains the above current and power in excess of what an ideal motor would take at same speed.

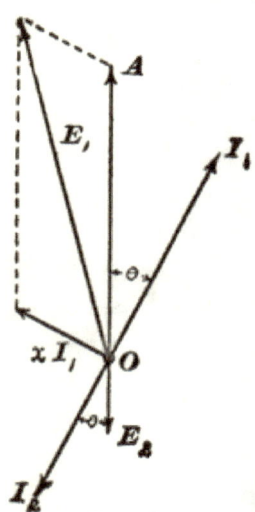

Fig. 181.

Further, there is eddy current and hysteresis loss in the rotor iron when its runs below synchronism and the effect of this loss is to slightly increase the torque.

145. Effect of stator resistance upon the action of an induction motor.—When the rotor is running nearly in synchronism with the stator magnetism the currents in the stator windings are very small and no perceptible portion of the supply e. m. f. is needed to overcome the stator resistance. As the rotor is slowed up the stator currents increase and a larger and larger portion of the supply e. m. f. is needed to overcome the stator resistance. The result is that the core flux* falls off slightly, and also the torque acting upon the rotor, inasmuch as this torque depends upon both flux and rotor currents.

146. Effect of magnetic leakage upon the action of an induction motor.—As in case of the simple transformer the effect of magnetic leakage is the same as the effect of an outside inductance connected in series with the primary (stator) windings, a separate

---

* Inasmuch as the portion of the supply e. m. f. which is balanced by the induced e. m. f. in the stator windings is decreased, and, therefore, the harmonically varying flux which induces this e. m. f. must decrease exactly as in the simple transformer.

inductance for each stator circuit. Let P be the value of each inductance and let x ($= \omega P$) be its reactance value.

The diagram of Fig. 181 represents the action of an induction motor in so far as it is affected by magnetic leakage; A and E_2 ($= sA$) are the e. m. f.'s induced in stator and rotor windings respectively by the magnetic flux which passes through both. The current I_2 is determined by the resistance and reactance* of the secondary circuit. The primary current I_1 is equal and opposite to I_2. The line xI_1 at right angles to I_1 represents that part of the total primary e. m. f. which is used to overcome the leakage inductance P or to balance the e. m. f. induced in each primary circuit by the leakage flux.

Let R be the resistance of each rotor circuit (inductance negligible).

Since I_1 and I_2 are equal we may represent both by the symbol I. The secondary current is $\frac{E_2}{R}$ or since $E_2 = sA$ we have:

$$I = \frac{sA}{R}. \qquad (a)$$

When the angle θ is zero then xI_1, Fig. 181, is at right angles to A and we have

$$E_1^2 = A^2 + x^2 I^2$$

or using the value of I from equation (a) and solving for A^2 we get

$$A^2 = \frac{E_1^2 R^2}{R^2 + s^2 x^2}. \qquad (b)$$

The power intake of primary, per circuit, is $P' = AI$ since A is the component of E_1 parallel to I. Therefore substituting for I its value from (a) and for A^2 its value from (b) we have:

$$P' = \frac{sE_1^2 R}{R^2 + s^2 x^2}. \qquad (c)$$

---

*Under practical conditions the rotor circuits are noninductive and the angle θ, Fig. 181, is zero, as in case of the simple transformer feeding a noninductive receiving circuit.

The electrical power output per secondary circuit is $P'' = E_2 I$. Therefore substituting sA for E_2, substituting for I its value from (a), and substituting for A^2 its value from (b) we have:

$$P'' = \frac{s^2 E_1^2 R}{R^2 + s^2 x^2} \qquad (d)$$

The mechanical power output per phase is $M = P' - P''$ or:

$$M = (1-s)\frac{s_1 E_1^2 R}{R^2 + s^2 x^2} \qquad (e)$$

The torque, T, acting upon the rotor, per phase, is such that

$$M = 2\pi n' T. \qquad (f)$$

in which n' is the rotor speed.

Let f be the frequency of the supply currents, p the number of pairs of poles for which the motor is wound, and n the speed of the stator-magnetism, then

$$f = pn. \qquad (g)$$

Let $s \left(= \frac{n - n'}{n} \right)$ be the slip, then $n' = (1-s) n$, or using the value of n from equation (g), we have

$$n' = \frac{f}{p}(1-s). \qquad (h)$$

Substituting the value of M from (e) and the value of n' from (h), in equation (f) we have

$$T = \frac{p}{2\pi f} \cdot \frac{s E_1^2 R}{R^2 + s^2 x^2}.$$

This is the torque per phase and since there are q phases the total torque is:

$$T = \frac{pq}{2\pi f} \frac{s E_1^2 R}{R^2 + s^2 x^2}. \qquad (110)$$

This is the equation to the curve, Fig. 182, of which the ordinates represent the torque T and the abscissas represent the slip s positive to the left and negative to the right. This figure is essentially the same as Fig. 177.

Maximum torque.—The slip s corresponding to maximum torque is found from the condition

$$\frac{dT}{ds} = 0$$

which gives
$$s = \pm \frac{R}{x}. \qquad (111)$$

The value of the maximum torque is found by substituting this value of s in equation (110) which gives:

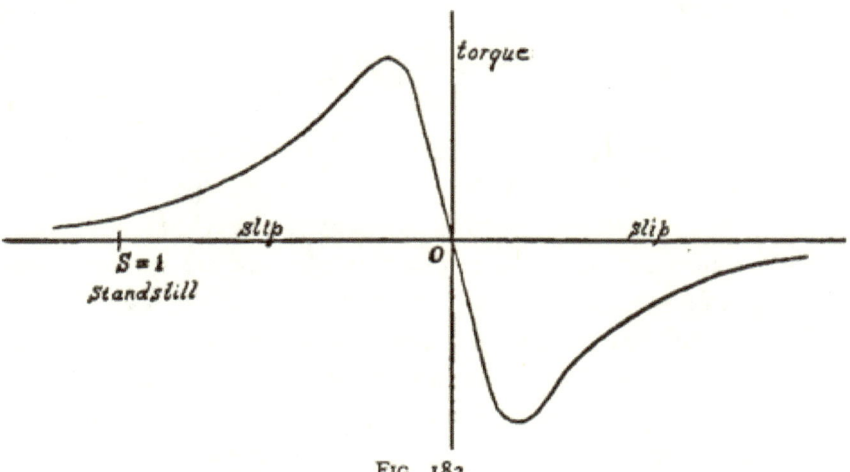

Fig. 182.

$$T_{max} = \frac{pq}{2\pi f} \cdot \frac{E_1^2}{2x}. \qquad (112)$$

The maximum torque is therefore independent of secondary resistance.

Starting torque.—When $s = 1$ the rotor is stationary and the corresponding value of T is the starting torque T_1, namely:

$$T_1 = \frac{pq}{2\pi f} \cdot \frac{E_1^2 R}{R^2 + x^2}. \qquad (113)$$

To give the greatest starting torque R must be adjusted to make T_1 a maximum. The condition is

$$\frac{dT_1}{dR} = 0,$$

which gives $R = x.$ (113)

That is, to give maximum torque at starting the rotor resistance per circuit must be equal to the leakage reactance per circuit.

147. Calculation of leakage reactance.—The leakage reactance x per circuit is equal to ωP ($= 2\pi f P$) where P is the leakage inductance per circuit. This leakage inductance is calculated, as in case of the simple transformer, by equation (75), namely

$$P = \frac{4\pi Z'^2 \lambda}{l}\left(\frac{X}{3} + \frac{Y}{3} + g\right). \quad (75)\text{ bis}$$

This equation gives P in centimeters, all dimensions being in centimeters. In this equation λ in the length, parallel to the shaft, of the rotor or stator; l is the sum of the widths of all the slots in which the windings of one stator circuit are wound; X is the depth of the stator slots; Y is the depth of the rotor slots; and g is the clearance space between stator and rotor. This equation assumes that stator and rotor slots are of the same width, that they are wound full of wire, and that the permeability of the iron lugs between the slots is very great so that reluctance of iron is negligible.

148. Outline of the design of an induction motor. *Design of primary.*—This member, which is usually the stator or stationary member in practice, is designed in a manner precisely similar to the design of a polyphase alternator armature as follows:

Value and frequency of supply e. m. f., speed of motor* and output are usually prescribed.

Number of poles follows at once from frequency and speed. Inner diameter of stator is fixed by speed of rotation and allowable peripheral speed. Length is then determined so as to radiate the internal losses and so on. In the calculation of flux,

---

* Which is practically the speed of the stator-magnetism.

$\frac{1}{4p}$ of the inner face of the primary member may be taken as the approximate area of pole face p being the number of pairs of poles. Windings are usually distributed in from 2 to 6 slots per pole per phase. The relation between primary e. m. f. per phase, turns T per phase, flux from one pole, and frequency is given by equation (59).

Design of secondary.—Length and diameter of secondary member is determined by length and diameter of primary member. The secondary should have about the same amount of copper as primary. Squirrel cage winding or ordinary distributed winding may be used. The latter is frequently employed when it is desired to insert resistance in rotor winding at starting. Number of slots in secondary is usually different from number in primary to avoid simultaneous coincidence of all primary and secondary slots.

149. The action of the polyphase alternator as an induction motor when being started as a synchronous motor.—The winding of the polyphase armature is identical to the stator or primary winding of an induction motor. When, at starting, the armature is connected to the polyphase supply mains a rotating state of magnetism is set up in the armature core. This rotating magnetism exerts a dragging torque on the field magnet, especially if the field coils are short-circuited, and the reacting torque of the field upon the armature sets the latter rotating in a direction opposite to the direction of its rotating magnetism.

CHAPTER XV.

TRANSMISSION LINES.

150. Introductory.—Power may be transmitted by the pumping of water. If great pressure is used a given amount of power may be transmitted by a small flow of water through a small pipe. In every case, however, there is a loss of power on account of friction in the pipe. The smaller the pipe the greater this loss and the less the first cost; the best size of pipe is that for which neither the first cost nor the continuous loss of power by friction is excessive.

Similarly a given amount of power may be transmitted by a small electric current through a small wire by using a large electrical pressure or e. m. f. In every case, however, there is a loss of power on account of the resistance of the wire. The smaller the wire the greater this loss and the less the first cost of the line; the best size of wire is that for which neither the first cost nor the continuous loss of power by resistance is excessive.

It is only by using very large e. m. f.'s that long distance transmission lines may be made at a reasonable cost, the loss due to resistance being at the same time reasonably small. The highest e. m. f. that can be satisfactorily used upon a pole line exposed to the air is about 40,000 volts, inasmuch as the leakage from wire to wire (outgoing and returning wires) in the form of brush or spark discharge becomes excessive at about 60,000 volts unless the wires are very large and very far apart. For transmission within a radius of two or three miles 1000 and 2000 volts are usually employed.

151. Power and e. m. f. loss in line.—If, say, 10% of the power output of a direct-current dynamo is lost in the line, then 10% of the e. m. f. of the dynamo is also lost in the line and 90% only is effective at the receiving circuit. With alternating current, however, the receiving circuit may receive, say, 90% of the power output of the dynamo, while the effective e. m. f. at the receiving circuit may be more or less than 90% of the e. m. f. of the dynamo. The difference (numerical) between dynamo e. m. f. and the e. m. f. at the receiver circuit is called the *line drop* and this line drop is of more practical importance than the power lost in the line, inasmuch as nearly all receiving apparatus needs to be supplied with current at approximately constant e. m. f. This is usually provided for by over-compounding the dynamo so as to keep the receiver e. m. f. constant. Thus, if the line is designed to give 10% drop, the dynamo would be 10% over-compounded.

Problem VII., Chapter VII., involves the general question of line drop and the results of this problem are applied to the calculation of transmission lines to give a prescribed drop. This problem may be, however, considerably simplified for practical use as is shown later.

152. Line resistance.—The resistance of a wire for alternating currents may in all practical cases be taken to be the same as the resistance of the same wire for direct current. The fact is, however, that the alternating current near the axis of a wire lags in phase behind the current near the surface of the wire, and the resistance of the wire is therefore larger for an alternating current than for a direct current.

153. Line reactance.—The reactance of a transmission line (outgoing and returning wires side by side) is greater the smaller the wires and the further they are apart, and is proportional to the length of the line and to the frequency. The accompanying table gives the resistance and reactance per half mile of transmission line.

RESISTANCE AND REACTANCE OF ONE MILE OF WIRE (½ MILE OF TRANSMISSION LINE) (EMMET).

Size of wire B. & S gauge.	Resistance in ohms.	Reactance in Ohms.					
		At 60 cycles per sec.			At 125 cycles per sec.		
		Wires 12 inches apart.	Wires 18 inches apart.	Wires 24 inches apart.	Wires 12 inches apart.	Wires 18 inches apart.	Wires 24 inches apart.
0000	.259	.508	.557	.591	1.06	1.17	1.23
000	.324	.523	.573	.607	1.09	1.20	1.26
00	.412	.534	.588	.618	1.12	1.23	1.29
0	.519	.550	.603	.633	1.15	1.26	1.32
1	.655	.565	.614	.648	1.18	1.28	1.35
2	.826	.580	.629	.663	1.21	1.31	1.38
3	1.041	.591	.644	.674	1.24	1.34	1.41
4	1.313	.606	.656	.690	1.26	1.37	1.44
5	1.656	.620	.670	.704	1.30	1.40	1.47
6	2.088	.633	.685	.720	1.32	1.43	1.49
7	2.633	.647	.700	.730	1.35	1 46	1.52
8	3.320	.662	.712	.742	1.38	1.48	1.55
9	4.186	.677	.727	.761	1 41	1.51	1.58
10	5.280	.688	.742	.776	1.44	1.54	1.62

154. Line capacity.—The two wires of a transmission line constitute a condenser, and they are repeatedly charged and discharged with the pulsations of e. m. f. The current (charging current) which the lines take from the generator to charge and discharge them is approximately 90° ahead of the e. m. f. in phase, and if the major part of the line current is behind the e. m. f. in phase, as it usually is, the effect of the charging current is to slightly lessen the total current taken from the generator. This charging current is usually ignored in the calculation of transmission lines. The student will find a discussion of its effects in Chapter XII. of Steinmetz's "Alternating current phenomena."

155. Interference of separate transmission lines.—When more than one transmission line (more than two wires) is strung on the same poles the alternating current in each line induces e. m. f.'s in the other lines and affects the line drop. This interference of one line upon another is obviated by crossing the lines at every second or third pole as shown in Figs. 183, 184 and 185. Fig. 183 shows the arrangement of a single-phase alternating current line to avoid inductive effects upon any other line that may be in the neighborhood; Fig. 184 shows the arrangement of four wires for transmitting two-phase currents; and Fig. 185 shows the arrangement of three wires for transmitting three-phase currents.

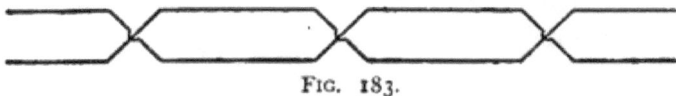

FIG. 183.

Remark: Transmission lines also affect neighboring lines by charging and discharging them electrostatically with the pulsations of e. m. f.; and by leakage currents due to incomplete insulation.

156. Calculations of a transmission line to give a specified line drop (*single-phase*).—A transmission line is usually designed to deliver a prescribed amount of power P at prescribed e. m. f. E

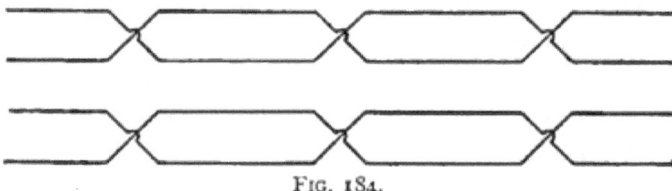

FIG. 184.

to a receiver circuit of which the power factor, $\cos \theta$ [see Art. 48], is given. The line drop, frequency, length of line and distance apart of wires are also given.

The generator e. m. f. E_0 is equal to the sum (numerical sum) of E and line drop.

The full load current I is found from $EI \cos \theta = P$.

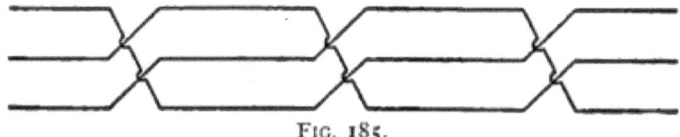

FIG. 185.

The component of E parallel to I is $E\cos \theta$, and the component of E perpendicular to I is $E \sin \theta$.

By treating the problem at first as a direct current problem the approximate resistance r' of the line is found, namely, $r'I =$ line drop. From this approximate resistance and length of line the approximate size of wire and line reactance x is found from the table; and since the line reactance varies but little with size of wire the value of x need not be further approximated.

The component of E_0 parallel to I is $E\cos\theta + rI$ where r is the true resistance of the line, and the component of E_0 perpendicular to I is $E\sin\theta + xI$. Therefore

$$E_0^2 = (E\cos\theta + rI)^2 + (E\sin\theta + xI)^2$$

or
$$r = \frac{\sqrt{E_0^2 - (E\sin\theta + xI)^2} - E\cos\theta}{I} \qquad (114)$$

From this equation the true line resistance r may be found and thence the correct size of wire.

Example:

  $E = 20,000$ volts,
  $P = 1,000$ kilowatts,
  $\cos\theta = .85 =$ power factor of receiving circuit,
  $E_0 = 23,000$ volts or line drop $= 3,000$ volts,
  frequency $= 60$ cycles per second,
  distance $= 30$ miles,
  distance apart of wires $= 18$ inches.

From these data we find:

  $I = 58.8$ amperes.
  $r' = 51$ ohms.

Therefore, from the table we find that, approximately, a No. 2 B. & S. wire is required so that $x = 37.7$ ohms.

Further
$$E\cos\theta = 17,000 \text{ volts,}$$
$$E\sin\theta + xI = 12,700 \text{ volts,}$$

and from equation (114) we find

$$r = 37.3 \text{ ohms}$$

from which the correct size of wire is found to be approximately a No. 1 B. & S.

157. Calculation of double line for two-phase transmission (*four wires*).—In this case each line is calculated to deliver half the prescribed power. Thus, if it is desired to deliver 1,000 kilowatts at 20,000 volts two-phase, at a frequency of 60, line drop of 3,000 volts, etc., then each line is calculated as a single-phase line to deliver 500 kilowatts at 3,000 volts line drop. The lines being, of course, arranged as shown in Fig. 184.

158. Calculation of a three-wire transmission line for three-phase currents.—The calculation will be carried out for the case of Y-connected generator and Y-connected receiver as shown in Fig. 186, for the reason that the relation between E_0, E, and line current is then the simplest.

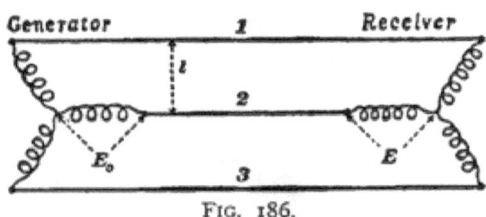

Fig. 186.

Let $\cos \theta$ be the power factor of each receiving circuit, P the total power to be delivered, E the e. m. f. between the terminals of each receiving circuit and E_0 the e. m. f. of each armature winding on the generator*; all prescribed.

Then
$$P = 3EI\cos \theta$$
from which the full load line current I may be calculated.

The difference $E_0 - E$ is due to e. m. f. drop in one main. Therefore, looking upon the problem as one in direct currents, we have $E_0 - E = r'I$ where r' is the approximate resistance of one main. From this the approximate size of wire may be found from the table.

Consider one of the mains; say, main No. 2; the other two mains together constitute the return circuit for this main, and the average distance from main 2 to mains 1 and 3 is $1\frac{1}{3}l$ when the mains are crossed as shown in Fig. 185. Find the reactance x of a pair of mains each of the size approximated above and distant $1\frac{1}{3}l$ from each other.

---

* The e. m. f. between mains at receiving station is $\sqrt{3}\ E$ and the e. m. f. between mains at generating station is $\sqrt{3}\ E_0$.

The component of E parallel to I is $E\cos\theta$ and the component of E perpendicular to I is $E\sin\theta$.

The resistance drop in one main is rI and the reactance drop in one main is $\frac{1}{2}xI$, the former being parallel to I and the latter perpendicular to I.

Then the components of E_0 are $E\cos\theta + rI$ and $E\sin\theta + \frac{1}{2}xI$ so that

$$E_0^2 = (E\cos\theta + rI)^2 + (E\sin\theta + \tfrac{1}{2}xI)^2,$$

or
$$r = \frac{\sqrt{E_0^2 - (E\sin\theta + \tfrac{1}{2}xI)^2} - E\cos\theta}{I}, \qquad (115)$$

which gives r, the true resistance of one main, from which the correct size of wire is easily found.

The calculation of a transmission line when the e. m. f.'s between mains is specified instead of the e. m. f.'s in Y-connected circuits is sufficiently explained in the following example.

Example: E. m. f. between mains at receiving station to be 20,000 volts. Therefore e. m. f. between terminals of Y-connected receiving circuits would be $20,000 \div \sqrt{3}$. Therefore

$$E = 11,550 \text{ volts}.$$

E. m. f. between mains at generating station to be 23,000 volts. Therefore $E_0 = 23,000 \div \sqrt{3}$ or:

$$E_0 = 13,280 \text{ volts}.$$

Further specifications:

$P = 1,000$ kilowatts,
$\cos\theta = .85$,
frequency $= 60$ cycles per second,
distance $= 30$ miles,
distance apart of adjacent wires $= 15\frac{3}{4}$ inches $(= l)$.

From these data we find

$$I = 34.0 \text{ amperes},$$
$$r' = 50.9 \text{ ohms}.$$

Therefore, approximately, a No. 5 wire is needed. The reactance, x, of a 30-mile double line of No. 5 wires at 21 inches ($= 1\tfrac{1}{3}l$) apart is, from the table,

$$x = 41.2 \text{ ohms.}$$

Equation (115) then gives

$$r = 46.5 \text{ ohms.}$$

So that a wire between No. 4 and No. 5 would give the prescribed line drop.

THE END.

INDEX.

Absolute electrometer, the, 33.
Admittance, definition of, 67.
Air gap, magnetic densities in, 113.
All-day efficiency of the transformer, 139.
Alternating currents, advantages of, 80.
Alternating current, utilization of, for motive purposes, 178.
Alternator, armature of, 19.
Alternator, brushes of, 19.
Alternator, characteristic curve of, 98.
Alternator, collecting rings of, 19.
Alternator, the constant current, 98.
Alternator, exciter of, 19.
Alternator, field magnet of, 18.
Alternator, fundamental equation of, 24.
Alternator, number of poles, 104.
Alternator output, limits of, 101.
Alternator, phase constant of, 100.
Alternator, polyphase, as an induction motor, 197.
Alternator, the simple, 18.
Alternator, speed and frequency of, 19.
Alternator speeds, 104.
Alternator, the single-phase, 80.
Alternator, the two-phase, 81.
Alternator, the three-phase, 84.
Alternators, compounding of, 116.
Alternators, design of, 114.
Alternators, field excitation of, 115.
Alternators, in parallel, 160.
Alternators in series, 151, 160.
Ammeter, the electrodynamometer, 29.
Ammeter, the hot wire, 29.
Ammeter, the plunger type, 33.
Analogies, mechanical and electrical, 15.
Armature conductor, current densities in, 113.
Armature core, magnetic densities in, 113.

Armature current, magnetizing action of, 95.
Armature current, reaction of, 95.
Armature drop, 97.
Armature inductance, 96.
Armature reaction, 95.
Armature of simple alternator, 19.
Armature windings, 105.
Armature windings, single-phase, 106.
Armature windings, two-phase, 107.
Armature windings, three-phase, 108.
Armatures, disk, 104.
Armatures, drum, 104.
Armatures, insulation of, 112.
Armatures, ring, 105.
Average values of e. m. f. and current, 23.
Average value of harmonic e. m. f. and current, 47.
Average values, definition of, 46.

Bar winding, 111.

Cardew's voltmeter, 29.
Characteristic curve of alternator, 98.
Collecting rings of alternator, 19.
Complex quantity, addition and subtraction of, 63.
Complex quantity, definition of, 62.
Complex quantity, multiplication and division of, 63.
Complex quantity, the use of, 62.
Composition and resolution of harmonic e. m. f.'s and currents, 42.
Compounding of alternators, 116.
Condenser, the, 14.
Conductance, definition of, 67.
Constant current alternator, 98.

Constant current transformer, 135.
Contact maker, the, 26.
Converter, the rotary, 166.
Converter, armature current of, 170.
Converter, armature reaction of, 170.
Converter, armature heating of, 175.
Converter, current relations of, 174.
Converter, e. m. f. relations of, 172.
Converter, hunting of, or pumping of, 169.
Converter, operation of, 168.
Converter, rating of, 171.
Converter, starting of, 168.
Converter, single-phase, 166.
Converter, two phase, 166.
Converter, three-phase, 166.
Converter, use of, 167.
Copper losses of transformer, 138.
Core flux, maximum value of, 123.
Core reluctance, effect of, on action of transformer, 124.
Coulomb, definition of, 13.
Current and e. m. f. curves, 20.
Current densities, 113.
Curves of e. m. f. and current, 20.
Cycle, definition of, 41.

Decay and growth of current, 9.
Decaying oscillatory circuit, 56.
Definition of admittance, 67.
Definition of average or mean value, 46.
Definition of complex quantity, 62.
Definition of conductance, 67.
Definition of the coulomb, 13.
Definition of cycle, 41.
Definition of electric charge, 13.
Definition of electrostatic capacity, 14.
Definition of the farad, 14.
Definition of frequency, 41.
Definition of harmonic current, 39.
Definition of harmonic e. m. f., 39.
Definition of the henry, 4.
Definition of impedance, 66.
Definition of inductance, 3.
Definition of magnetic flux, 1.
Definition of the microfarad, 14.
Definition of mutual inductance, 12.

Definition of opposition, 41.
Definition of period, 41.
Definition of phase constant, 100.
Definition of phase difference, 41.
Definition of power factor, 50.
Definition of quadrature, 41.
Definition of reactance, 66.
Definition of resistance, 66.
Definition of simple quantity, 62.
Definition of slip of induction motor, 189.
Definition of susceptance, 67.
Definition of synchronism, 41.
Definition of vector quantity, 63.
Delta connection for three-phase systems, 87, 91.
Design of alternators, 114.
Design of induction motor, 196.
Design of transformers, 140.
Design of transmission lines, 201.
Distortion of field by armature current, 95.
Distributed winding, effect of, 99.

Eddy current loss in iron cores, 137.
Eddy currents, effect of, on action of a transformer, 124.
Effective values of e. m. f.'s and currents, 23.
Effective values of harmonic e. m. f.'s and currents, 48.
Efficiency of alternators, 140.
Efficiency of induction motor, 185.
Efficiency of synchronous motor, 155, 157.
Efficiency of transformers, 138.
Electrical and mechanical analogies, 15.
Electric charge, definition of, 13.
Electric charge, measurement of, 13.
Electric resonance, 57.
Electrodynamometer, the, 29.
Electrometer, the absolute, 33.
Electromotive force and current curves, 20.
Electromotive force drop in transmission lines, 198.
Electromotive force, induced, 2.

INDEX.

Electromotive force, lost in armature, 97.
Electromotive force, self-induced, 7.
Electrostatic capacity, definition of, 14.
Electrostatic voltmeter, the, 32.
Equation, fundamental, of the alternator, 24.
Equivalent resistance and reactance of transformer, 122
Excitation characteristics of synchronous motor, 163.
Exciter of alternator, 19.

Farad, the, definition of, 14.
Field excitation of alternators, 115.
Field magnet of alternator, 18.
Flux, magnetic, definition of, 1.
Form factor of alternating e. m. f., 49.
Frequency and speed, relation of, 19.
Frequency changer, the, 189.
Frequency, definition of, 41.
Frequencies of alternators, 103.

Graphical method, 62.
Growth and decay of current, 9.

Harmonic current, definition of, 39.
Harmonic currents, composition and resolution of, 42.
Harmonic e. m. f., definition of, 39.
Harmonic e. m. f.'s, composition and resolution of, 42.
Harmonic e. m. f.'s and currents, average values of, 47.
Harmonic e. m. f.'s and currents, effective values of, 48.
Harmonic e. m. f.'s and currents, rates of change of, 44.
Henry, the definition of, 4.
Hot wire ammeter and voltmeter, 29.
Hysteresis, effect of, on action of a transformer, 124.
Hysteresis loss in iron cores, 137.

Impedance, definition of, 66.
Induced electromotive force, 2.
Inductance, calculation of, 8.

Inductance, definition of, 3.
Inductance, mechanical analogue of, 5.
Inductance of armature, 96.
Induction generator, 186.
Induction motor, the, 178.
Induction motor, action of, as a transformer, 188.
Induction motor, calculation of leakage inductance of, 196.
Induction motor, effect of eddy currents and hysteresis, 191.
Induction motor, effect of magnetic leakage, 192.
Induction motor, effect of speed upon transformer action of, 188.
Induction motor, effect of stator resistance, 192.
Induction motor, efficiency and rotor resistance, 185.
Induction motor, efficiency and speed, 185.
Induction motor, general theory of, 137.
Induction motor, maximum torque of, 195.
Induction motor, maximum starting torque, 196.
Induction motor, polyphase alternator as an, 197.
Induction motor, single-phase driving, 186.
Induction motor, slip of, 189.
Induction motor, starting torque of, 195.
Induction motor, stator windings of, 179.
Induction motor, the actual, 191.
Induction motor, the ideal, 190.
Induction motor, torque and speed, 184.
Induction motor used as a frequency changer, 189.
Induction motor, use of starting resistance, 184.
Induction motors, designing of, 196.
Induction motors, use of two, for street railway work, 187.
Insulation of armatures, 112.
Iron losses of transformer, 137.

Kinetic energy of the electric current 3.

INDEX.

Leakage current of transformer, 125.
Leakage inductance of a transformer, 131.
Leakage, magnetic, effect of, upon transformer, 139.
Line drop, 198.

Magnetizing current of transformer, 125.
Magnetic densities in armature and air gap, 113.
Magnetic densities for transformer cores, 141.
Magnetic flux, definition of, 1.
Magnetic flux, through a coil, 2.
Magnetic leakage, effect of, upon transformer, 130.
Mean values, definition of, 46.
Measurement of electric charge, 13.
Measurement of power, 35.
Mechanical and electrical analogies, 15.
Mechanical resonance, 59.
Mesh connection for three-phase system, 87. 91.
Microfarad, the, definition of, 14.
Moment of inertia, analogue of inductance, 5.
Monocyclic system, 148.
Motor, induction. See induction motor.
Motor, the synchronous. See synchronous motor.
Mutual inductance, definition of, 12.

Noninductive circuits, 4.

Opposition, definition of, 41.
Oscillatory current, the, 56.
Oscillatory current, the decaying, 56.
Output, influence of inductance upon, 103.
Output, limits of, of alternator, 101.
Output, limit of transformer, 139.

Period, definition of, 41.
Phase constant, definition of, 100.
Phase constants, table of, 101.
Phase difference, definition of, 41.
Plunger type, ammeter and voltmeter, 33.

Primary of transformer, 119.
Problem I, 9.
Problem II, 9.
Problem III, 51.
Problem IV, 51, 66.
Problem V, 53.
Problem VI, 54, 66.
Problem VII, 69.
Problem VIII, 73.
Problem IX, 75.
Problem X, 76.
Poles of alternators, number of, 104.
Power developed by harmonic e. m. f., 49.
Power factor, definition of, 50.
Power, instantaneous and average values of, 22.
Power in polyphase systems, 91.
Power, measurement of, 35.
Power measurement, three-ammeter method, 36.
Power measurement, three-voltmeter method, 35.

Quadrature, definition of, 41.

Rates of change of harmonic e. m. f.'s and currents, 44.
Rating of transformers, 140.
Ratio of transformation of transformer, 119.
Reactance, definition of, 66.
Reaction of armature currents, 95.
Recording wattmeter, the, 38.
Rectifying commutator, 116.
Regulation of the transformer, 127.
Regulation of the transformer, calculation of, 134.
Resistance, definition of, 66.
Resolution and composition of harmonic e. m. f.'s and currents, 42.
Resonance, electric, 57.
Resonance, mechanical, 59.
Rotary converter, the. See converter.
Rotor, the squirrel cage, 179.
Rotor, definition of, 179.

INDEX.

Scott's transformer, 148.
Secondary of transformer, 119.
Self-induced electromotive force, 7.
Simple quantity, definition of, 62.
Single-phase armature windings, 106.
Slip of induction motor, 189.
Spark gauge, the, 33.
Speed and frequency, relation of, 19.
Speeds of alternators, 104.
Squirrel cage rotor, 179.
Stator, definition of, 178.
Stator windings of induction motor, 179.
Steinmetz' method, 62.
Steinmetz system of tandem control for induction motors, 187.
Susceptance, definition of, 67.
Symbolic method, application of, 66.
Synchronism, definition of, 41.
Synchronous motor, the, 151.
Synchronous motor, excitation characteristic of, 163.
Synchronous motor, fundamental equations of, 154.
Synchronous motor, greatest e. m. f. of, 161.
Synchronous motor, greatest intake of, 161, 162.
Synchronous motor, hunting of, or pumping of, 169.
Synchronous motor, stability of, 159.
Synchronous motor, starting of, 157.

Table of line resistance and reactance, 200.
Table magnetic densities for transformer cores, 141.
Table of phase constants, 101.
Table of transformer efficiencies, 138.
Thomson inclined coil ammeter and voltmeter, 34.
Thomson recording wattmeter, 38.
Three-ammeter method, 36.
Three-phase armature windings, 108.
Three-phase alternator, 84.
Three-phase e. m. f.'s and currents, 83, 85.

Three-phase system, balanced, 90.
Three-phase system, e. m. f. and current relations, 88.
Three phase system, power, 91, 93.
Three-voltmeter method, 35.
Transformation, ratio of, of transformer, 119.
Transformer, the actual, 124.
Transformer, the ideal, 119.
Transformer connections, 143.
Transformer, constant current, 135.
Transformer cores, magnetic densities for, 141.
Transformer design, 140.
Transformer, effect of coil resistances, 129.
Transformer, effect of core reluctance and hysteresis upon, 124.
Transformer, effect of magnetic leakage, 130.
Transformer efficiency, 138.
Transformer efficiency, all day, 139.
Transformer losses, 137.
Transformer, magnetizing or leakage current of, 125.
Transformer output, limits of, 139.
Transformer, primary and secondary of, 119.
Transformer rating, 140.
Transformer, ratio of transformation, 119.
Transformer regulation, 127.
Transformer regulation, calculation of, 134.
Transformer, Scott's, 148.
Transformer, the general alternating current, 187.
Transformer, two-phase, three-phase, 146
Transformers with divided coils, 144.
Transmission lines, 198.
Transmission line capacity, 200.
Transmission line, e. m. f. drop, 198.
Transmission line reactance, 199.
Transmission line resistance, 199.
Transmission lines, designing of, 201.
Transmission lines, interference of, 200.
Trigonometrical method, 62.
Two-phase armature windings, 107.

Two-phase alternator, 81.
Two-phase system e. m. f. and current relations, 84.
Two-phase system balanced, 84.
Two-phase system, power, 91.

Units of electric charge, 13.
Units of inductance, 4.

Vector, addition and subtraction, 64.
Vector division, 65
Vector multiplication, 64.
Vector, numerical value of, 64.
Vector quantity, definition of, 63.

Voltmeter, the absolute electrostatic, 33.
Voltmeter, Cardew's, 29.
Voltmeter, the electrodynamometer, 31.
Voltmeter, the electrostatic, 32.
Voltmeter, the hot wire, 29.
Voltmeter, the plunger type, 33.
Voltmeter, the spark gauge, 33.

Wattmeter, the, 36.
Wattmeter, the recording, 38.
Windings of armatures, 105.

Y-connection for three-phase system, 87, 90.

www.ingramcontent.com/pod-product-compliance
Lightning Source LLC
Chambersburg PA
CBHW031831230426
43669CB00009B/1301